CAD FRAMEWORKS

PRINCIPLES AND ARCHITECTURE

THE KLUWER INTERNATIONAL SERIES IN ENGINEERING AND COMPUTER SCIENCE

VLSI, COMPUTER ARCHITECTURE AND DIGITAL SIGNAL PROCESSING

Consulting Editor

Jonathan Allen

CAD FRAMEWORKS

PRINCIPLES AND ARCHITECTURE

by

Pieter van der Wolf

*Department of Electrical Engineering,
Delft University of Technology,
Delft, The Netherlands*

KLUWER ACADEMIC PUBLISHERS

BOSTON / DORDRECHT / LONDON

A C.I.P. Catalogue record for this book is available from the Library of Congress

ISBN 0-7923-9501-8

Published by Kluwer Academic Publishers,
P.O. Box 17, 3300 AA Dordrecht, The Netherlands.

Kluwer Academic Publishers incorporates
the publishing programmes of
D. Reidel, Martinus Nijhoff, Dr W. Junk and MTP Press.

Sold and distributed in the U.S.A. and Canada
by Kluwer Academic Publishers,
101 Philip Drive, Norwell, MA 02061, U.S.A.

In all other countries, sold and distributed
by Kluwer Academic Publishers Group,
P.O. Box 322, 3300 AH Dordrecht, The Netherlands.

Printed on acid-free paper

Printed in the Netherlands

CONTENTS

PREFACE

Since the early 1980's, the topic of CAD frameworks has received a great deal of interest, both in the research community and in the commercial arena. It is generally agreed upon that CAD framework technology holds a great promise: Advanced CAD frameworks can turn collections of individual tools into effective and user-friendly design environments. How can this promise be fulfilled?

This book gives an in depth discussion of the design and construction of CAD frameworks. It presents principles for building integrated design environments and shows how a CAD framework can be based on these principles. In a systematic way it derives the architecture of a CAD framework, using well-defined primitives for representation. This architecture defines how the many different framework sub-topics, ranging from e.g. concurrency control to design flow management, relate to each other and come together into an overall system.

The origin of this work is the research and development performed in the context of the Nelsis CAD Framework. For the last eight years, the Nelsis CAD Framework has been a working system, gaining functionality while evolving from one release to the next. The principles and concepts presented in this book have been field-tested in the Nelsis CAD Framework.

This book is primarily intended for EDA professionals, both in industry and in academia, but is also valuable outside the domain of electronic design. Many of the principles and concepts presented are also applicable to other design-oriented application domains, such as mechanical design or computer-aided software engineering (CASE).

The book includes an index and an extensive bibliography. The contents is organized as follows.

Chapter 1 stresses the need for CAD frameworks and identifies causes for the slow progress in the framework area.

Chapter 2 investigates the state of the art in the area of CAD frameworks, and their architectures in particular. It presents the principal requirements a CAD framework has to satisfy.

Chapter 3 starts the architecture definition process with a global investigation into the nature of a CAD framework. It defines a global framework model based on principles for data management and tool - framework interaction.

A framework architecture can be viewed from many perspectives, ranging from end-user views to implementation views, each one emphasizing particular aspects of the architecture. This book follows a three step approach to the presentation of the framework architecture, these steps being:

1. Description of the information architecture.

2. Description of the component architecture.

3. Description of the implementation architecture.

Chapter 4 discusses data modeling techniques and introduces a semantic data model for communicating and documenting the information architecture.

Chapter 5 employs the semantic data model to derive the information architecture in the form of a data schema. This data schema represents the logical organization of the design environment in terms of object types and their relationships. It provides a basis for discussing framework functionality as well as for structuring information in the framework upon implementation.

Chapter 6 presents the component architecture of the CAD framework. It identifies the individual framework components and their relationships. Relevant interfaces of the framework components are defined. Special attention is paid to the tool - framework interface. The characteristics of this interface determine whether integrated design environments can be built successfully on top of the CAD framework. Chapter 6 also discusses the framework user interface, which offers facilities for browsing the design information and the design history and permits tools to be launched conveniently. These facilities demonstrate that a CAD framework can indeed be the designer's powerful and friendly assistant.

The implementation architecture presented in chapter 7 describes how the framework is implemented in terms of Unix operating system primitives.

The design choices made at this level are decisive for obtaining maximum efficiency and optimal behavior with respect to physical distribution and multi-user support.

Chapter 8 presents concluding remarks.

Acknowledgements

Many colleagues and students have contributed to the ideas presented in this book. I would like to thank everyone who has been involved in the design and construction of the Nelsis CAD Framework, in particular Teus Vogel, Nick van der Meijs, Alfred van der Hoeven, Anthon Ouwendijk, Jos Hegge, Gijs Sloof, Simon de Graaf, Ing Widya, Peter van Putte, Peter Kist, Mattie Sim, Cees Schot, Wim Tiwon, and Olav ten Bosch. Special thanks go to Peter Bingley for the fruitful cooperation over the years and the great brainstorm sessions. I would also like to thank Rene van Leuken for directing and managing the Nelsis team. I am grateful to Prof. Patrick Dewilde for his guidance and inspiration and for creating the research environment in which this work could be born. Further, I would like to thank everyone who commented on earlier drafts of this text, in particular Sally Kleinfeldt and Peter van den Hamer.

The development of the Nelsis CAD Framework was supported by the Nelsis program, funded partially by the Dutch Government, and the ICD program which was funded partially by the Commission of the EC under the MR09 and Esprit 991 programs. Thereafter, CAD framework research at Delft University of Technology has been funded partially by the Commission of the EC under the Esprit 5082 and Esprit 7364 programs (JESSI-Common-Frame). The support is gratefully acknowledged.

Delft, July 1994   Pieter van der Wolf

1

INTRODUCTION

1.1   THE NEED FOR CAD FRAMEWORKS

More and more these days, the bottleneck in the development of advanced electronic products is *design*. More and more designs of increasing complexity have to be done faster and faster to bring more advanced end-products to the market in time [Man92, dG93]. The key to gains in design productivity is Electronic Design Automation (EDA). The optimistic EDA picture is that a large number of *tools* have become widely available. For specific design tasks these tools help the designer to master the complexity and perform these tasks efficiently. Together with the interactiveness provided by modern graphical workstations this has yielded significant productivity improvements for parts of the design process.

Due to this focus on *tool automation*, design systems have become large tool-boxes offering the designer a great variety of loosely coupled tools to perform the many design tasks. The realistic EDA picture, however, is that these tools support only individual design tasks, leaving the designer with the problem of handling the multitude of tools and of successfully moving his design data from one tool to the other. Moreover, the number of tools to be operated is growing continuously, as are the amounts of design data to be handled. What is lacking is *integration* and overall support for managing the *design process*.

People have recognized these problems and have realized that attention has to be paid to the overall efficiency of the design process in order to continue achieving gains in productivity. As a consequence, one of the buzzwords in the EDA arena is *CAD framework*. CAD stands for Computer-Aided Design.

1

We adopt the following definition of CAD framework, as originally given by the CAD Framework Initiative (CFI), the international consortium developing framework standards [CFI90b]:

Definition 1.1 (CFI) *A* CAD framework *is a software infrastructure that provides a common operating environment for CAD tools.*

CAD frameworks play a role in *building* as well as in *operating* integrated design environments. First, a CAD framework has to provide facilities for conveniently integrating multiple CAD tools into a coherent design environment. It is a basis for *tool integration*. In this respect there is a direct analogy with the role an operating system plays in the development of general-purpose software applications.

Second, a CAD framework can support the end-user in conveniently operating the design environment. Being the infrastructure that binds the tools together, the framework is the proper place to incorporate facilities for organizing the design information and managing the design process. In an integrated design environment, these facilities can support the end-user in keeping track of the state of his design and in applying tools effectively. This will help him to master the complexity of the design and the design process. More complex circuits, satisfying more stringent performance and quality standards, can then be done faster under the increasing time to market pressures. Thus, a CAD framework is to become the *electronic assistant of the designer* for organizing the design information and managing the design process.

From the above description we identify two categories of framework users: *developers* (e.g. CAD tool developers, CAD tool integrators) and *end-users* (e.g. design engineers, administrators, project managers). To get a more concrete idea of what the introduction of CAD frameworks may yield in practice for the end-user, consider the following support that a CAD framework is expected to give:

- Help the design engineer to maintain an overview of his design descriptions: which components are available, what is their status and history, how are the components related.

- Tell the design engineer which tools are available for which design tasks and give information on their usage. Moreover, a framework may take care that tools are applied in the proper sequence such that a pre-defined design methodology is adhered to.

■ Make design projects manageable by performing book-keeping on the status of achievement for consultation by the project manager.

■ Allow teams of design engineers to cooperate effectively on a design project. A related buzzword here is *concurrent engineering*, which is "the art of decomposing a complex serial task into smaller, relatively independent tasks that can be executed in parallel" [CFI90a].

Effective framework technology providing such advanced facilities as exemplified above can thus help to:

■ *Cut down design time.* Recent studies show that for every six months a project is late, the potential profit is reduced by a third [DEC92].

■ *Make the design process less error prone.* According to recent studies, design errors account for an average of 20 percent of product costs [DEC92].

■ *Master the increasing complexity* of the design and the design process.

■ *Increase performance and quality* of electronic products.

In addition, a CAD framework provides *environmental stability* as it offers a standard operating environment to an ever evolving set of CAD tools, in which many services have yet been captured. It promotes *modularization* of CAD systems, as these are to be constructed as cooperating tool components on top of a CAD framework, rather than being implemented as monolithic super-tools. In fact, CAD tools may become simpler as the generic services of CAD frameworks become more powerful. Further, we think that CAD frameworks do not simply add to what a designer has to learn; Increased uniformity and user-friendliness may actually make CAD systems easier to operate.

1.2   THE SEARCH FOR CAD FRAMEWORKS

There appears to be world-wide consensus on the point that there is an urgent need for CAD framework technology in order to build effective integrated design environments that help improve design productivity. Also, the EDA community appears to agree on the major functional requirements for a CAD framework. However, there is no common idea how to go about developing and implementing one. As is said in [Val90], "The tremendous amount of importance placed on frameworks, as they directly relate to improving productivity, combined with the lack of agreement on their exact nature has shrouded the subject in mystery and uncertainty". The current situation is that effective framework based design systems satisfying the major framework requirements have not yet reached the designer's workbench [vdH91, Mal92]. In fact, major EDA vendors are investing heavily but have problems in providing the required functionality while making the system efficient [BHNS92].

The CAD Framework Initiative (CFI), the international consortium developing framework standards, is trying to alleviate the mystery surrounding frameworks. The mission of CFI is: "To develop industry acceptable guidelines for design automation frameworks which will enable the coexistence and cooperation of a variety of tools" [CFI90b]. Having standard interfaces to CAD framework components will substantially reduce the cost of combining multi-sourced CAD tools into an effective design environment. But, progress is slow and consensus on specifications, detailed requirements and implementations appears difficult to reach [vdH91].

The major European framework effort is the Jessi-Common-Frame (JCF) project, which is aimed at building a CAD framework. Also in this project the different development partners, though having extensive framework experience, appeared to have quite different views on how to build a framework. Extensive exertions were required in order to identify critical differences in the respective approaches and to obtain a common understanding on a global framework architecture.

Now, what makes it so difficult to develop, implement, and even discuss CAD frameworks? Obviously CAD frameworks are complex systems; this is the nature of the subject. First of all frameworks have to satisfy many functional and operational requirements posed by different categories of users. A great variety of services has to be provided to tool integrators, design engineers, administrators, etc. A CAD framework is not a piece of software performing a specific design task in a specific way with measurable results. In many respects it must be generic, customizable, and open, in order not to have built-in restrictions for a particular usage of the system. For example, it should not make restrictive assumptions on the kinds of tools to be supported. The generic nature required for many framework functions adds to the complexity. Further, the framework is a multi-user system, which must provide services to many concurrent users in a distributed hardware environment, responding well under all kinds of circumstances. This significantly adds to the complexity of the framework problem.

In addition to the complex nature of CAD frameworks, discussions on framework architectures are hampered by the lack of agreement on a common formal "language" or "model" to represent and discuss framework architectures. Each active member of the framework community appears to have his/her own way of (informally) representing certain aspects of a framework architecture. As a result, ideas on framework architectures do not get communicated well. It is hard for framework developers to disengage from their own background and learn the essence of the work of other people. A confusion of tongues in framework discussions is often the result.

In the following chapters we respond to the situation sketched above by systematically deriving the architecture of a CAD framework. This architecture will be described by a global framework model and three more specific framework views. We will define key principles to direct design choices and will use well-defined primitives to describe the framework views.

Two key features of the CAD framework architecture will be *openness* and *efficiency*. Openness is the ability of the CAD framework to easily incorporate new tools, new types of data, and new design methodologies. It must be flexible, configurable, and largely independent of a specific application domain. Efficiency implies that overall efficiency of the design process is optimized. This relates to run-time performance of individual framework services as well as to framework functions that help the end-user to work more effectively.

Before we start the architecture definition process, we investigate in the next chapter the state of the art in the area of CAD frameworks, and their architectures in particular. We also present the principal requirements a CAD framework has to satisfy.

2

STATE OF THE ART AND REQUIREMENTS

2.1 THE EVOLUTION OF CAD FRAMEWORKS

2.1.1 File-and-Translator Based Systems

From the literature it appears that over the last ten years people have gradually discovered the topic of CAD frameworks, and have gradually allocated more functional requirements to this topic. The state of the art in CAD frameworks is the result of increasing awareness and evolutionary developments rather than specific scientific breakthroughs. We therefore start our assessment of the state of the art from an historical perspective.

The history of CAD frameworks starts in the early eighties when people realized that with the growing number of CAD tools data transfer between tools became an ever growing pain. At that time tools typically used proprietary and tool-dependent ascii or binary formats to represent the design data. Communication between design tools was possible only if a translator for the respective formats was available [Kat83, Kal85, Kat85b, GE87, HNSB90]. This situation is illustrated in Figure 2.1. Bringing the growing number of tools together into an integrated design environment involved writing a large number of translators. The emergence of de-facto standard formats, allowing the number of translators to be reduced, eased this pain a bit.

These tool integration efforts had made people realize that effective EDA solutions not only had to provide the individual tools but also the integration facilities, or 'glue', to support the communication between tools. The integration facilities in the file-and-translator based systems consisted of translators

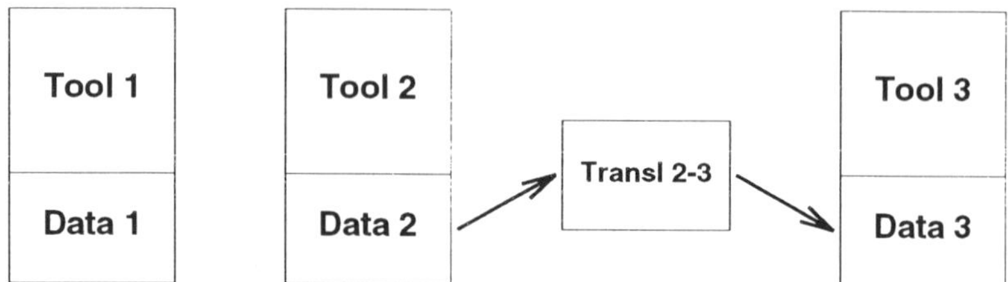

Figure 2.1 Tools having proprietary data representations. Translation across data representations is required for sharing data among tools.

and a whole collection of ad-hoc utilities to make life easier for the end-user. These simple integration facilities can be termed the first primitive CAD frameworks, and a new EDA topic had been born.

2.1.2 First Role: Design Database

Historically, the first role allotted to CAD frameworks is that of *common data repository* [NPSVtS81, Kat82, Goe85, Tur85, CFHL86]. Data which is common to a number of CAD tools is stored only once in the repository, from where it can be used as input for all tools. For tools that have not been written to operate directly on the common data repository, reformatting may be performed on input and output. The common repository has come to be called 'integrated design database', or just '*design database*'. It is the key to *tool integration*, and in particular to *tool interoperability*: the ability of CAD tools to communicate and share data. See Figure 2.2.

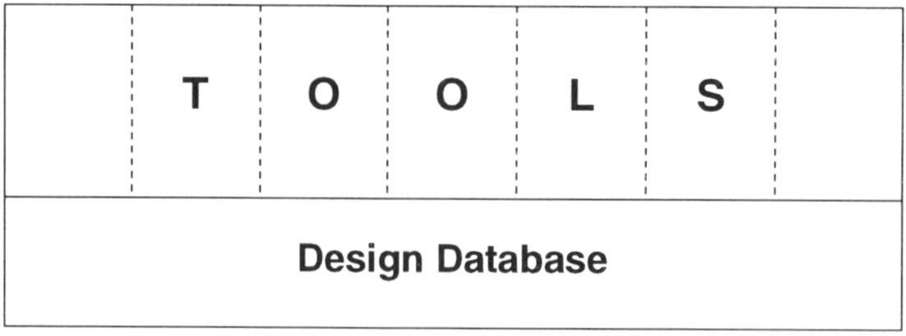

Figure 2.2 Tools integrated on top of a design database.

Advantages of this approach are increased consistency and increased efficiency. For the end-user there is increased convenience as his design data is stored in a structured way in a single place. Additional database utilities support the end-user in managing his design data.

File Based Systems

In many CAD systems the design database is a file based system, implemented as a small layer on top of the host operating system [NPSVtS81, Kat83, Kat85b, May87]. This layer augments the operating system with specific support for managing the design files. Design files are stored in a structured file organization, for example, employing a hierarchical file system and pre-defined naming conventions. Access methods are provided for storage and retrieval of design data. These access methods may offer atomic update capabilities.

Examples of the file based approach are the Nelsis CAD Framework, Release 2 and Release 3 systems [Dew86]. Several more advanced CAD frameworks still employ file based systems for persistent design data storage. Examples are Design Framework II from Cadence Design Systems, Falcon from Mentor Graphics, PowerFrame from Digital Equipment Corporation, and the Nelsis CAD Framework, Release 4.

Use of Conventional DBMSs

Another approach has been to employ conventional (record-oriented) database management systems (DBMSs) to fulfil the role of design database. These conventional DBMSs are typically used as storage and transaction processing components in business applications. The interest in DBMSs was driven mainly by the apparent 'keyword-level' match between DBMS functions and facilities and the emerging requirements for design databases. A DBMS provides mechanisms for the reliable storage of data, including recovery facilities, it protects data from unauthorized access, and it provides concurrency control and integrity maintenance. The notion of *transaction* supported by these systems also seems useful: a sequence of database operations that are either executed completely or not at all [Dat86].

Quite a number of attempts to employ conventional database techniques have been published in the literature [Hay81, RBJ81, Zin81, Kat82, WBS82, Hay83, HNCL84, Har84]. However, none of them reported a straightforward applicability of off-the-shelf DBMSs. Most of the conventional DBMSs have been targeted for business applications and do not specifically address the problems encountered in a design environment. Critical differences occur in characteristics of data to be handled (e.g. granularity and inter-relatedness) and access characteristics (e.g. frequency, granularity and duration of transactions) [Sid80, LP83, Buc84, HY85, SA86]. There is a mismatch between the facilities provided by the DBMSs and the requirements posed by engineering applications. Due to this mismatch, efficiency is hard to obtain, if at all, given the internal mechanisms (for concurrency control and recovery, for example), which have been geared towards the characteristics of business applications. We are currently not aware of any successful application of a classical business DBMS for the purpose of a CAD framework design database, and consider this to be a dead end.

Successes have been reported in the application of conventional DBMSs for meta data handling only, that is, to hold administrative data including references to the actual design data stored in files. An example of the use of a conventional DBMS for meta data handling is [BD91].

Object-Oriented DBMSs

The database community has recognized the fundamental nature of the requirements posed by engineering applications and other new database applications such as multimedia databases and knowledge bases. Work was started on next-generation DBMSs targeted at these applications: *object-oriented DBMSs*. In contrast to conventional DBMSs, object-oriented DBMSs offer far more flexibility for handling highly interrelated data of different granularities on which different types of access are performed. For engineering applications, the DBMS must be capable of handling many different data types and large numbers of instances of each type. Moreover, flexibility must be provided for modifying or extending the conceptual schemas to match the different views of data operated on by different application programs [SA86, BP86].

Databases typically found in Artificial Intelligence (AI) systems tend to provide flexibility for many different data types, which may also be defined dynamically. An example of such a system is the Pearl AI Package (Package for Efficient Access to Representations in Lisp) [DFW82]. These systems, however, have not been geared towards the large numbers of instances of each type, as found in engineering applications [SA86]. Moreover, they typically are single-user systems. They may be used for purposes of prototyping, but are not suited as production systems.

The data models provided by powerful object-oriented DBMSs permit arbitrary types of design information to be represented and accessed conveniently. Typically the design information is modeled as a web of highly *interrelated objects* [HMSN86, FFHe91, Obj91, HJR92]. Granularity of the objects may vary significantly; from a simple integer attribute value to a complete design file. Typically, a *traversal type of access* is provided to the application programs built on top of the DBMS. This permits them to navigate through the object web, to retrieve objects or to add new ones to the web. A key in achieving efficiency for EDA applications is maintaining locality of larger aggregates of data that are typically accessed as a (compound) entity. For this purpose object-oriented DBMSs targeted at these applications provide such features as *clustering* or *complex objects* [LP83, Buc84, HS87, FFHe91, Obj91, HPC93].

A number of object-oriented DBMSs have been realized and have become the base components of some of today's frameworks. For example, the Objectivity/DB [Obj91] has been used for meta data management in ValidFrame [Val90] and the Cadlab OMS (Object Management System) [FFHe91] is used in the Jessi-Common-Framework [JCF91b]. Cadence Design Systems has announced an agreement with Object Design, Inc. to use the ObjectStore DBMS in its next-generation EDA software. These developments signal a move by the EDA software industry away from proprietary design databases to the use of standard commercially available DBMSs.

2.1.3 Second Role: Design Data Manager

The second major role allotted to CAD frameworks is that of *design data manager*. A design database, in the sense of common data repository, provides a facility for storing the design data, but provides no support for *managing* the data. Randy Katz from U.C. Berkeley was one of the first to realize that 'brains' could be added to the design database 'muscle', to actually help the designer in organizing his design information [Kat83]. A design data management system uses knowledge of the structure and status of design information to provide management support and enforce constraints on the design process.

Clearly, system integration can be, and should be, much more than the definition of common formats for the purpose of tool communication. With the tools being interfaced to the framework, common data management services can be identified and incorporated into the framework. Such data management services may, for example, organize design information across representations, provide versioning capabilities to support evolutionary design, support consistent operation on hierarchical designs, control concurrent access and facilitate teamwork. A user interface may then be added to enable the end-user to interact with the system to get informed about the structure and status of his design. "Which copy is the latest version?", "Has this layout been extracted since it was updated, and if so, which circuit description was derived from it?", "If I change this layout, which other parts of the design will be affected?". It is the ability to answer such questions that differentiates a true data management system from a simple data repository [NSV86].

Since the mid 1980's the research community has been relatively active in the data management area. No real breakthroughs in this area can be reported, though. Many publications address interesting problems in too much isolation, yielding partial solutions that do not fit in a wider, generally accepted, context. No generally agreed upon set of data management principles has been established. Common agreement on *some* aspects of data management is reflected by the terminology that has been developed. For example, some commonly used terms are *meta data*, *design object*, and *design transaction*. At present there is a growing awareness that an overall concept is needed in which the individual data management topics are harmonized, and that formal means such as *information modeling* should be applied to develop such a concept.

Today, data management facilities can be found in several framework based CAD systems. Some examples are Falcon from Mentor Graphics and Design Framework II from Cadence Design Systems. Other CAD framework products offering data management functionality are PowerFrame from Digital Equipment Corporation and SiFrame from SNI. The Nelsis CAD Framework, Release 4, offers data management functionality based on integral meta data management [vdWSBD90].

Many of the data management solutions presented in today's CAD systems appear to be ad-hoc extensions rather than a coherent set of facilities based on a well-defined conceptual foundation. This is illustrated by missing features or serious restrictions found in these systems. For example, in some systems design hierarchies can be traversed only in the downward direction or design hierarchies have to be isomorphic across representations. Data management solutions adding value to today's CAD systems do exist, but too often they lack important functionality, are too slow, or miss required flexibility.

2.1.4   Third Role: Design Process Manager

The third major role allotted to CAD frameworks is that of *design process manager*. With the increasing number of tools found in today's CAD systems, there is a growing need to support the design engineer in correctly executing these tools to perform his design tasks. Framework services may help the design engineer in correctly invoking the individual tools, as well as provide support for executing the tools in the correct order, according to a pre-defined design procedure. A next step is to have the framework automatically execute design tools, for example, when valid output data is required for a subsequent tool run or the verification status of the design is to be enhanced. Ultimately, design process management facilities will help offer a design environment in which the framework actively supports the design engineer in meeting his design goals.

In the late 1980's the research community 'discovered' the topic of design process management. A pioneering publication was [Jan86]. Key research has been performed at Carnegie Mellon University [BD89, BD91], U.C. Berkeley [CNSVl90], Philips Research Laboratories [vdHT90], Siemens [BKL+90], and Delft University of Technology [tBBvdW91]. In [KGMB94] Kleinfeldt et. al. define the important concepts in this area and give an extensive overview of the state of the art.

The terms *design methodology management* and *design flow management* are both used to denote framework services that help the design engineer to correctly perform design activities according to a pre-defined design procedure. The term *design methodology* is typically used to refer to the definition of a design procedure in terms of abstract design stages, such as 'circuit design', 'circuit verification', and 'layout design'. The term *design flow* typically refers to the definition of a design procedure in terms of individual tools and dependencies between tools.

Design flow management is one of the key framework topics today. As demonstrated by several prototypes [Jan86, tBBvdW91], a pre-defined design flow can be presented graphically to the end-user, to actually guide him through the design process. On some aspects of design flow management there appears to be quite natural agreement: a design flow can be represented by a graph describing how data can flow from one tool to the other, and validity of data can be indicated in such a graph with appropriate primitives. At a more detailed level, however, the known approaches differ significantly in their capabilities; for example, in the types of tools that can be supported and in the restrictions put on data management functions. The prototype implementations also differ significantly from an architectural point of view. For example, some systems provide design flow management through a special framework tool, while other systems consider this functionality to be located in a framework kernel component [BtBvdW92]. This will be further addressed in chapter 6.

The limited design flow management capabilities found in today's CAD systems, if present at all, are by no means the full-fledged services they should be. In the commercial arena customers are aware of the potential advantages and CAD vendors are under pressure to provide advanced services that help the end-user perform his design tasks. Design flow management functionality is starting to appear in commercial products. Examples are recent framework products from Cadence Design Systems and Viewlogic Systems and SiFrame from SNI. The Nelsis CAD Framework, Release 4.5, includes a design flow manager which guarantees correct use of data and execution of tools in the correct order, according to a configurable design flow description. It has been implemented as a kernel framework service for full multi-user capabilities and support for interactive tools. A graphical flow browser enables the designer to inspect the status of his design and to invoke tools [tBBvdW91, BtBvdW92, tBvdWB93].

Many more developments can be expected in the area of design process management. Design flow management is now primarily concerned with correct use of data and the execution of tools in the correct order, but higher level facilities for design methodology management and design task scheduling will certainly follow. Further, more emphasis will be put on design decision support services, which help design engineers in their decision-making processes, for example, by estimating the costs related to alternative design solutions and checking these against the design objectives. New developments are also expected in the area of *project management*. These will include computerized facilities for the planning of projects, selection of tools and methodologies, allocation of tasks to team-members, evaluation and presentation of project progress, etc.

2.2 STATE OF THE ART
IN FRAMEWORK ARCHITECTURES

2.2.1 Introduction

We now take a more specific look into the state of the art in framework architectures. In Webster's English dictionary [Neu88] the 'architecture' of 'something' is defined as its 'design and construction'. We consider the *architecture of a CAD framework* to be a description of its principal functions and global structure as seen from the outside, as well as a description of the principal mechanics of its inner structure. It defines *what* functions are provided by the framework (specification side) as well as *how* these are provided (implementation side). Given the specification side, the implementation side gives additional detail on the construction of the framework, which is crucial to satisfying requirements on physical distribution, multi-user support, and run-time efficiency.

Looking into the literature it becomes clear that little research explicitly addresses the architecture of CAD frameworks including aspects of their construction. It appears that there is far less knowledge and consensus on how to build a framework than on what it should do. Also there appears to be no generally accepted formalism in which to present (views on) framework architectures. When architectural issues are addressed in a publication, typically a single architectural view of a system (for example, its communication architecture) is presented through informal drawings, thus being incomplete and leaving much space for misinterpretation. Below we present an overview of relevant literature on (the constructive aspects of) framework architectures, seeking both for important architectural concepts and for ways of representing these concepts.

2.2.2 The Research Community

U.C. Berkeley, Oct/VEM

A well known system is the Oct/VEM facility for data management and graphics editing, developed at U.C. Berkeley [HMSN86]. The Oct data manager provides a basis for tool integration by acting as a central repository for data interchange. The VEM graphics editor provides browsing and editing capabilities for Oct designs. In order to be flexible, Oct makes very few assumptions about the data to be handled. The high level object types are *cell*, *view*, and *facet*, where a cell contains views, and a view contains facets. Facets are described in terms of basic objects, e.g. boxes, polygons, etc, related through *attachment*. This information model employed by Oct is presented in [HMSN86] in an informal way. The interface for design data access consists of a set of procedures allowing applications to open facets and iterate over their contents. The internal structure of Oct and VEM is not further detailed in [HMSN86]. Aspects of communication between application programs and Oct/VEM, in particular with respect to operation in a distributed hardware environment, are described and exemplified.

In [SGKN89], Silva et. al. describe extensions to Oct/VEM, named Octane. Facilities for versioning, configuration management and concurrency control are proposed. Again, there is no formal presentation of the (extended) information model. The software structure is presented in terms of coarse Oct- and Octane modules and their relationships.

U.C. Berkeley, Version Server

In [Kat83] and following publications [KW84], [Kat85a], [Kat86], [KAC86], [KBC⁺87], Randy Katz from U.C. Berkeley presents a system for design data management. He concentrates on coarse-grain management of design data, rather than focusing on handling of detailed design data elements. Management is performed at the level of *design objects*, being convenient aggregations of design information. Both an information model and an overview of the system components are presented. The information model, called "data structures for specifying a design", defines the structure for organizing *alternatives*, *versions*, and *representations* of a design as well as *hierarchical* and *equivalence* relationships. Presentation of the information model is done in an informal way through intuitive drawings and a textual description. The layered ordering of the system components is given, and their functioning is described textually. There is no description of internal interfaces or a tool interface. No implementation architecture is presented.

'Electronic CAD Frameworks', Tutorial Paper / Book

In the tutorial paper [HNSB90] and book [BHNS92] entitled "Electronic CAD Frameworks", Harrison et. al. present an extensive tutorial on CAD frameworks. They present an overview of the history of CAD frameworks, describe the state of the art for the key framework topics, and identify directions for further research, development and standardization. The major components of a CAD framework are identified and ordered, to reflect the functional decomposition of a CAD framework. The authors stress that an incremental approach is to be adopted to the design and implementation of a CAD framework.

The Pace Framework

In [dSSV90] and [SdS91] the Pace framework, developed at Inesc, Portugal, is described. It includes subsystems offering database, user interface, and intertool communication services. Spook is a base subsystem offering mechanisms for interprocess communication and for storage of raw data on files. Ghost is the framework user interface subsystem. Dwarf is the database subsystem offering Oct-like data modeling capabilities.

Spook, Ghost and Dwarf are general-purpose base components that allow tools to be tightly integrated. Emphasis is on message-based communication between applications. The Pace components do not yet provide higher level data or design management services.

Philips Research, The Roadmap Model

In [vdHT90] van den Hamer and Treffers from Philips Research Laboratories present the Roadmap model. This data flow based model addresses task definition, flow definition, and tool execution. A key feature of the Roadmap model is the integration of data management and design flow management. In a design environment, the configured design flow is used as the basis for organizing and accessing design data. The design flow defines the logical structure for storing the design derivation history. A flow-based user interface permits intuitive browsing of design data and design derivation history through interaction with the design flow.

Van den Hamer et. al. use information modeling techniques to formally represent the information architecture of their system in terms of object types and relationships between object types. They do not identify (functional) components of the system and do not present an implementation architecture.

Carnegie Mellon University

Prominent research activities in the area of CAD frameworks have been performed at Carnegie Mellon University (CMU) under the direction of Stephen Director. The main focus of the work has been on the topics of tool management and design process management.

The Ulysses system [BD89] is an early approach to design process management based on a blackboard model and the application of artificial intelligence techniques. Ulysses supports automated execution of sequences of CAD tools in a goal-directed fashion. Design space exploration is supported by facilities that help restore design consistency upon design changes or violation of constraints. The Ulysses architecture is described in [BD89] in terms of the main system components and their relationships. Cadweld [DD89] is a follow on to Ulysses. By applying object-oriented techniques, it enhances the capabilities to describe, classify, and interact with CAD tools.

The Odyssey CAD framework includes a tool and data encapsulation component (Cyclops), a task and flow management component (Hercules) [BD91], and a planning component (Minerva) [JD92]. Hercules allows an application environment to be modeled by a *task schema*, a dependency graph that represents all possible task sequences. The task schema also specifies the logical structure for a task database that stores the design derivation history. This compares to the use of design flows in the Roadmap model [vdHT90].

The Nelsis CAD Framework

The Nelsis CAD Framework has been developed at Delft University of Technology. It is the 'software output' of one of the main research activities performed in the framework area over the last decade. The evolution of the Nelsis CAD Framework shows a direct parallel with the subsequent allocation of framework roles presented in section 2.1. Starting from a very basic system for structured file management, it has grown to become a full-fledged system offering data management and design flow management functionality.

The Nelsis approach in doing framework research has *not* been to build multiple research prototypes, each one addressing an individual framework sub-topic. Rather, the ambition has been to build a single overall system in which the different framework sub-topics can be researched and experimented with, also in relationship to each other. This permits concepts for data and design management to be mutually harmonized, and implemented for validation in a real design environment. Following this approach, the Nelsis CAD Framework has been a working system gaining functionality while evolving from one release to the next. From this perspective the Nelsis CAD Framework can be termed *"a framework for research into CAD frameworks"*.

The Nelsis CAD Framework makes a clear distinction between the detailed design data contained in design objects and the meta data which is maintained by the framework [vdWvL88, vdWBD90]. Information modeling techniques have been applied to structure the meta data. Design data management and design flow management services have been built on top of an advanced meta data handling facility. The Nelsis CAD Framework demonstrates framework principles and functionality, such as:

- Data schema driven meta data management [vdWSBD90].

- A structured Data Management Interface (DMI) for tool integration [vdMvLvdW$^+$87].

- A graphical user interface for meta data browsing and manipulation [BvdW90, tBvdWB93].

- Design data management functions such as versioning, support for multi-view hierarchical design, consistency management, etc.

- Design flow management [tBBvdW91, BtBvdW92, tBvdWB93].

- Multi-user support [WvLvdW88].

- Physical and logical distribution [SBD$^+$90, vdWSBD90].

- Openness, Flexibility, Modularity, Portability.

- Performance.

The Nelsis CAD Framework has been the vehicle for validating many of the ideas presented in this book. It allowed the many aspects of a CAD framework architecture to be judged in a real-life environment, functionality to be implemented in an overall context, and (unknown) problems as well as possibilities to be discovered, considered, and taken as input for further research.

2.2.3   The Commercial Vendors

EDMS / PowerFrame

Around 1986 EDA Systems introduced the Electronic Design Management System (EDMS) [BG87, GS88] and several years later its successor PowerFrame[1] [Joh89]. EDMS can be considered the first CAD framework available on the market as a stand-alone product. As such it has played a pioneering role in creating awareness in the EDA community: CAD frameworks were for real!

EDMS and PowerFrame are general-purpose CAD frameworks, suited for integration of tools from disparate sources into a single CAD system. They offer data and design management services, including a graphical desktop shell for data browsing and tool invocation. The data management services include version management, hierarchical composition and configuration management, and view management. PowerFrame offers basic design flow management capabilities.

The data management services are based on an information model, representing the different types of relationships between design objects. This information model is presented in an informal way. Further, in [BG87] the individual framework components are identified and their operation is described. Aspects of communication between the different components are presented and several interfaces are identified.

Though there are some clear technical differences, the EDMS / PowerFrame approach is very much in line with the work of Katz mentioned above. Management is also performed at the coarse-grain level of design objects, basically corresponding to design files. Information about the design objects, e.g. their relationships and history, is administered. Upon integration, tools are allowed to retain their own storage structure for the actual design descriptions, for example, a dedicated file organization.

---

[1] Now a product of Digital Equipment Corporation.

ValidFrame

ValidFrame has been a product of Valid Logic Systems. This company merged with Cadence Design Systems, San Jose, California. Cadence plans to incorporate concepts and components from ValidFrame into its own framework developments. According to [Val90], ValidFrame does not enforce all tools to interface to one monolithic database. Instead they are allowed to maintain their optimized, application-specific (file based) databases. ValidFrame focuses on maintaining administrative data about the design and design process. The Objectivity/DB system, a commercially available object-oriented database management system, is used to handle this administrative data. ValidFrame's Design Manager is claimed to provide extensive functionality, but usability problems have been reported. ValidFrame basically lacks design flow management functionality.

Cadence / Digital, Model of Framework Functionality

In [BT90] and [Cad90], Cadence Design Systems and Digital Equipment Corporation present a unified model of framework functionality. The model is an attempt to distinguish between the various levels of functionality provided by frameworks. It can be used to evaluate, compare, and classify frameworks.

The model has seven layers, each one relating to a particular kind of functionality. In the lower layers we find base services and design representation services. Then come data management and tool management services. In the top most layers are the design process / design methodology management services. Notice the analogy with the subsequent allocation of different roles during the evolution of CAD frameworks, as presented in section 2.1.

Applying the model to the systems presented above, we conclude that the Oct/VEM system and the Pace framework focus on the lower layers of functionality. Randy Katz, the Roadmap model, the CMU work, Nelsis, EDMS / PowerFrame, and ValidFrame, on the other hand, aim at providing higher level data and design management services, leaving the design representation issues to the tool developer or tool integrator. Their approaches, therefore, relate to the higher layers of the model.

2.2.4 Cooperation Projects

The CAD Framework Initiative (CFI)

One of the most dominant efforts in the framework area is the CAD Framework Initiative (CFI), an international non-profit consortium of CAD tool users, tool vendors, and research institutions. The primary aim of CFI is to define standards and guidelines that will allow tools from different sources to cooperate in a single design environment. The standardization is driven by the understanding that no single commercial CAD vendor can deliver state-of-the-art tools for all stages of the design process and all application domains. CAD users urgently need the ability to easily "mix & match" CAD tools from different vendors as well as home-brewed tools, to comfortably cook their favorite design environment. The ability to easily incorporate a CAD tool into a design environment is also referred to as "plug and play". The situation today is that for every dollar spent on tools, two dollars are spent integrating them [DEC92]. To improve this situation, the interfaces between tools and integration platform, the CAD framework, have to be standardized.

CFI has identified individual areas for standardization, for example, Design Representation and Inter-Tool Communication. Working Groups (formerly Technical Subcommittees) have been installed to actively work on the definition of standards in these areas. The Architecture Working Group works on the definition of a framework architecture reference, in order to provide an overall context to which the contributions of the other Working Groups must relate. This framework architecture reference is presented in the working document of the Architecture Working Group [CFI91a, CFI93]. See also [CFI91b].

The framework architecture reference is described by a number of alternative *views* that provide additional insight into the structure and relationship of CAD framework components. The views distinguished by CFI in [CFI93] are:

- The *user view*. This view recognizes objects and functions using these objects visible to end-users and developers. Information modeling techniques are applied to represent the object types and their relationships. This is done at a very abstract level, in terms of such object types as user, data, resource, task, methodology, etc.

- The *tool integration view*. This view addresses the elements of a tool relevant to the framework. These include tool abstractions, tool executables, and tool data files.

- The *framework services view*. From a tool's perspective, a framework is a collection of interfaces to services needed by CAD tools. These services can be grouped logically into components. The framework services view identifies the individual components and defines the dependencies between these components.

- The *communications view*. This view provides additional detail by focusing on the communication among framework components and tools. It shows how data, control and notifications are passed between these.

- The *data management view*. Using information modeling techniques this view characterizes the domain independent information that the data management services of the framework maintain about the design objects.

- The *design information view*. This view describes the domain specific types of information that are used to describe the actual design.

The CFI framework architecture reference is still evolving. It does not yet present a definite detailed picture of a framework architecture.

An illustrative CFI framework view is a global component view named the framework backplane view, presented in [CFI91a]. See Figure 2.3 (Figure 10 of [CFI91a]). This backplane view has been widely accepted, in- and outside CFI, to convey the sense that from a tool perspective a framework is very much like an electronic product card cage. Individual CAD tools, like T2 in Figure 2.3, can be 'plugged' into a CAD framework, to be used in cooperation with the other CAD tools and available framework services. Replacing one tool for the other can be done simply by 'unplugging' the first one and 'plugging' in the second one.

The framework backplane view identifies major components of a CAD framework (shaded components in Figure 2.3). However, it is not intended to represent relationships between components. The framework backplane provides common services that are required by CAD tools. Tools that conform to the standard interfaces to these services can be inserted without change.

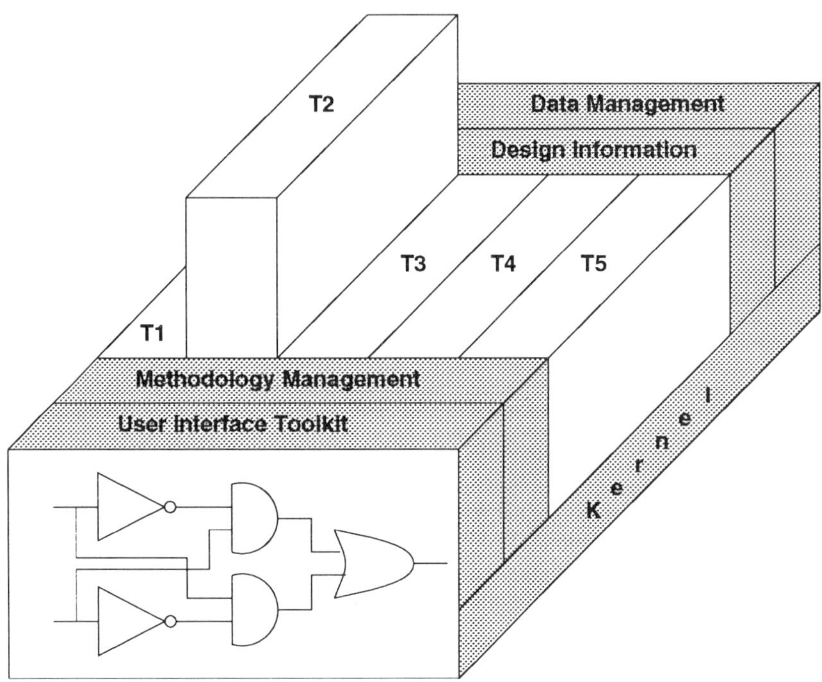

Figure 2.3   The CFI framework backplane view.

Another interesting CFI framework view is the framework services view, which identifies the individual components and the dependencies between these components. The CFI picture for this view is shown in Figure 2.4 (Figure 15 of [CFI93]).

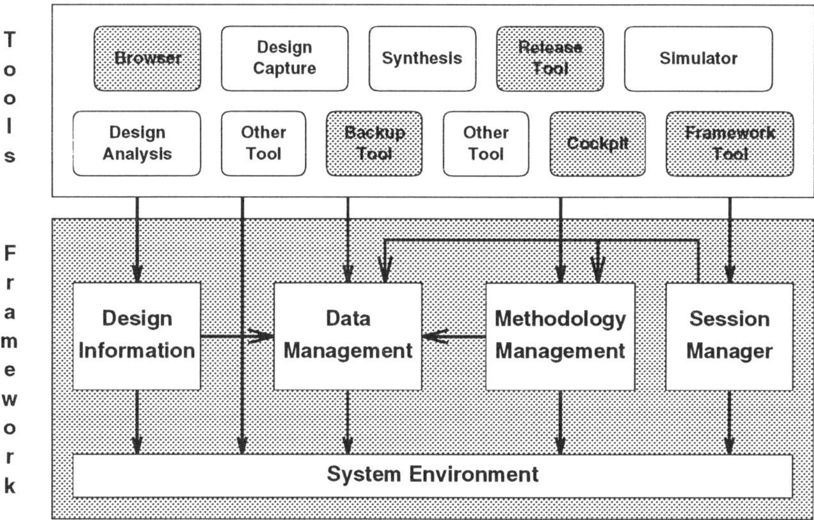

Figure 2.4   The CFI framework services view.

A number of logical components together form the framework. Each component has a set of programming interfaces that represents the services exported by that component, that is, made available to other components and CAD tools. Standards for the interfaces to CAD tools are the primary output of CFI.

- The basic component of the framework is the so called *system environment*. It provides the environment within which components and tools can run. These include portability services, which facilitate interchangeability of software across dissimilar hardware platforms, error handling services, and an extension language engine.

- The *design information* component provides programming interfaces that permit tools to create, modify, and access domain specific design data. For specific application domains, standard information models define the object types and relationships involved.

- The *data management* services are domain neutral services focusing on the interaction between tools and data. These services include version management, configuration management, access control, and relationship management.

- The *methodology management* component provides services that focus on the incorporation of tools and providing support for the design process in which these tools are to be used.

- The *session manager* supports a user during a session by maintaining a session context containing information about which tools are running, user information, and system resources.

Notice the correspondence between the framework components and the different roles allotted to CAD frameworks in the course of history, as sketched in section 2.1.

The arrows in Figure 2.4 show the dependencies among the framework components. Also shown is a collection of tools. All interaction with the end-user takes place through tools, run by the user. The framework components are passive; they wait for control to be passed from a tool. Some of the tools are *framework tools*, aimed specifically at interaction of the end-user with the framework, for example, to browse or manage design information.

Jessi-Common-Frame

Another major framework effort is the European Jessi-Common-Frame (JCF) project. This project was started in 1990 and involves many European companies and institutes, among which Siemens Nixdorf Informationssysteme AG (SNI), Cadlab, Philips, ICL, Delft University of Technology, GMD, and IMEC. The primary aim of CFI is the definition of standards. The JCF project, on the other hand, is actually building a framework, the Jessi-Common-Framework, which is intended to be compliant with the standards defined by CFI. The Jessi-Common-Framework is marketed by SNI under the name SiFrame.

The architecture for the Jessi-Common-Framework was specified by the project partners SNI, Cadlab, Telesoft and Delft University in 1991, and is described in [JCF91b]. A key view of the JCF architecture is the 'functional structure', depicted in Figure 2.5. Similar to the CFI framework services view, it identifies the logical components and the dependencies between these components.

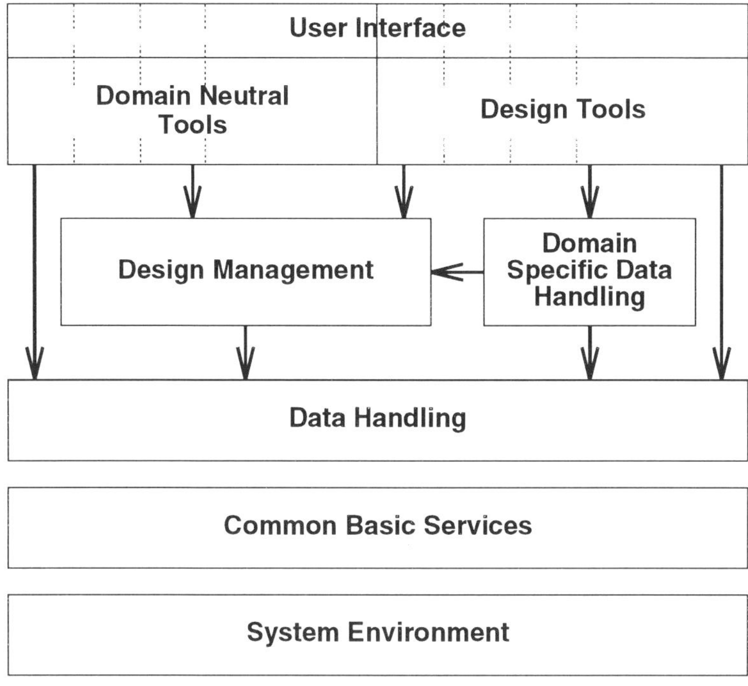

Figure 2.5   The Jessi-Common-Framework functional structure.

All interaction with the user is through tools, which are either design tools or domain neutral tools (framework tools). The tools rely heavily on the design management services for e.g. flow management, high-level consistency control, etc. The domain specific data handling offers comfortable access to standardized domain specific data. For example, the CFI Design Representation Procedural Interface (CFI DR-PI) [CFI92b] may be located here. All access to the unified database is controlled by the data handling component. This is an object-oriented DBMS providing the services for storing and retrieving data. It ensures consistency and security at low levels. The common basic services provide general services to the other framework components. Example services are error handling and a help system. The system environment provides a virtual operating system, the X Window System, OSF/Motif, etc.

In [JCF91b] the JCF architecture further describes the information flow between the functional blocks. It distinguishes between meta data and design data, where meta data contains information about the design data. The meta data is used for management purposes and is fully controlled by the design management component. The different types of objects managed by the framework are described. Semantic data modeling techniques are used to a limited extent to formally represent the object types and their relationships.

2.3 PRINCIPAL REQUIREMENTS

2.3.1 Introduction

We will now present the principal requirements a CAD framework has to satisfy. First we take a second look at the CFI definition of a CAD framework, which we have adopted in the introductory chapter, section 1.1: "A CAD framework is a software infrastructure that provides a common operating environment for CAD tools". It is important to notice that this definition in itself is very generic, as are the definitions in [HNSB90] and [JCF92]; it does not include a particular set of functions and features to be provided. This is due to the fact that the idea of what a CAD framework actually is, has gradually evolved over the last years and is expected to further evolve in the near future. Definitions which try to be specific about the functions and features to be provided by a CAD framework are bound to be outdated as engineering methods evolve, as both the software and hardware architectures of computer systems evolve, and as the needs and priorities of end-users change. As is said in [HNSB90], there is no "right answer" to the CAD framework problem. A good understanding of this fact is crucial for the approach one takes to the design of a CAD framework: key goals must be *configurability, flexibility, modularity* and all other keywords that imply easy adaptation and modification as requirements on CAD frameworks evolve.

In presenting the requirements, we make a distinction between

- required *overall operational characteristics*, and

- *functional requirements.*

As overall operational characteristics we consider required properties that are not so much related to individual framework functions or capabilities, but, rather, to the operation of the framework as a whole. The functional requirements relate to actual data and design management services to be provided to developers and end-users.

As functional requirements tell *what* (functions) the framework has to provide, the operational characteristics give constraints on *how* these functions are to be provided; the operational characteristics can be considered adjectives to the functional requirements. For example, in defining design flow management services, overall characteristics such as user friendliness, configurability, and performance have to be taken into account.

2.3.2   Overall Characteristics

1. *User friendly*:

   The framework must be the friendly assistant of the designer, rather than his dumb slave, his big brother watching him, or his severe master. Its operation must be comprehensible, it must prevent the designer from faulty actions without being too restrictive, and it must have a powerful user interface that is consistent and intuitive in appearance and behavior.

   As Knuth says in [Knu81]: *The enjoyment of the tools one works with is, of course, an essential ingredient of successful work.*

2. *Open*:

   The framework must be easy to deploy in the construction and actual use of an integrated design environment. It must be an open framework for tool integration that does not restrict the functionality of the design environment or force a design methodology on its users. With a minimum effort foreign tools must be allowed to operate as consistent parts of the integrated environment.

3. *Configurable*:

   Many aspects must be configurable from outside the system rather than being hardwired deep inside, in order to be able to satisfy different user requirements at different sites at different times and for different applications. This may range from configurability of user preferences at the user interface level to configurability of a design methodology that is to be supported by the framework in the course of a design project. Configurability can significantly contribute to openness (requirement 2).

4. *Flexible, extensible, modular, portable, maintainable*:

   The framework must be able to *evolve* smoothly to satisfy new requirements. This may involve adaptation or extension of existing framework services, extension with new framework services, porting to a new hardware platform or operating system release, maintenance, etc. Hence the framework must be well-structured, modular, with well-defined internal interfaces, rather than being an inflexible monolith.

5. *Performance*:

   The framework may offer fantastic high-level services for data and design management, with capabilities that go beyond the imagination of an ordinary human being; however, it will not be used if the performance is not adequate! Introduction of the framework facilities should cause no significant run-time performance degradation for the tools interacting with the framework. Response of the framework services with which the user interacts directly must be good.

The key goal is overall optimization of design efficiency, in particular for large complex designs. This refers to run-time efficiency of individual tools and services, as well as to increased framework functionality that helps the design engineer to work more effectively.

2.3.3 Functional Requirements

The following functional requirements of framework services are generally agreed upon [CFI90b, JCF90, HNSB90]:

1. *Facilitate tool integration*:

   The framework must allow convenient and efficient incorporation of design tools, to let them become consistent parts of the integrated design environment. It must facilitate *tool interoperability* by providing means for tools to communicate and share data through a common interface.

   To fulfil its role of design database, the framework must provide facilities for reliable persistent storage of design data. This involves data of a wide range of granularity for which the access characteristics may vary significantly.

2. *Support multiple users performing in parallel multiple design tasks on multiple design projects*:

   A *Logical distribution*:

   The framework must provide contexts in which design activities can be pursued without interaction with other (unrelated) design activities. We refer to this capability as *logical distribution* of design data and design activities. At any time it must, however, be possible to make results of one design activity available to others (design library facility).

   B *Multi-user, multi-tasking support*:

   Multiple users must be allowed to operate concurrently on a design project in a controlled and coordinated way. The framework must support cooperation between members of a design team, yet prevent interactions that may yield inconsistent design descriptions (see also requirement 4 below) or may otherwise be detrimental to design productivity.

Moreover, each single user must be allowed to perform multiple tasks in parallel, for example, run multiple tools from different windows on his graphics workstation. The framework must at all times guarantee consistency under concurrent operations.

3. *Physically distributed operation*:

The framework must operate in a distributed heterogeneous computing environment. It must support effective use of computing and storage facilities in this networked environment. Details of physical distribution should be made transparent to tools and end-users.

4. *Guarantee consistency and integrity of design information*:

The framework must prevent the design information from getting into an invalid state. This applies to the internals of individual design descriptions as well as to relationships between design descriptions. The process of change, inherent to evolutionary design, has to be managed to ensure the integrity and correctness of each portion of a design, and of the design as a whole.

This requirement relates to requirement 2B, in that consistency has to be maintained under concurrent operations, but also covers consistency maintenance that is to be performed in non-concurrent environments. For example, the framework must provide facilities for correct propagation of engineering changes throughout a design.

Note that this requirement relates only to consistency and integrity constraints that the framework is to guard. For example, the framework may consider a circuit description to be in a valid state, even though logic simulation has demonstrated it to have incorrect logic behavior.

5. *Access control*:

The framework has to check whether a certain user has permission to access certain design information to perform a certain operation (authorization). Access must be prohibited if the user does not have proper permissions.

6. *Support Evolutionary Design*:

Design is iterative and tentative. Design engineers often take their design descriptions through several *refinements* and explore different design *alternatives* seeking for the best solution.

To support evolutionary design, multiple versions of a design description must be allowed to co-exist. The framework must maintain a derivation history and support selection and use of individual versions. This framework service is typically referred to as *version management*.

7. *Support hierarchical multi-view design*:

Typically a design is described hierarchically at multiple levels of abstraction. The framework has to support the systematic management of hierarchical multi-view design information. It has to permit and promote coordinated (parallel) design activities being performed on the different components in the design hierarchy and at the different levels of abstraction.

8. *Configuration management*:

The system under design is typically decomposed into different components, with design efforts on these components being performed in parallel. The framework has to support the systematic management of collections of (versions of) components that somehow fit together. This service is typically referred to as *configuration management*. A *configuration* is a collection of related design objects, used for purposes such as release management and design checkpointing [JCF90].

9. *Tool management*:

The framework must help the design engineer in conveniently executing tools. For example, it may aid the design engineer in selecting the proper tool for a given task, in correctly invoking the tool, and in supervising running tools.

10. *Design flow management*:

The framework has to assist the design engineer in correctly and efficiently performing design activities according to a locally defined design procedure. It must inform the design engineer about the design state and the design history (what has been done), and support him in selecting a design action that he is allowed to take in a given design state (what can be done).

2.4 CONCLUSION

From the historical sketch presented in section 2.1, we conclude that the idea
of what a CAD framework is and which functionality it should provide has
gradually evolved. Starting from the design database as the basis for tool
integration, it has grown to the design data manager and design process manager that assists the design engineer in performing his complex design task.
This has been an evolutionary process of increasing awareness rather than a
revolution driven by certain technology leaps. We investigated the state of
the art in framework architectures and presented principal requirements for
a CAD framework. These principal requirements are quite generally agreed
upon. Detailed requirements, on the other hand, often raise a lot of discussion
and discord. The understanding that requirements on CAD frameworks are
still evolving and often depend on local preferences is of great importance
in a framework architecture definition process.

3

GLOBAL FRAMEWORK MODEL

3.1 INTRODUCTION

In this chapter we start the definition of the CAD framework architecture by making some initial observations and by defining some terms, principles and global concepts. This will lead to a global framework model. In subsequent chapters we will refine this global model by presenting different architectural views that describe different framework aspects at a more detailed level.

A framework based design environment is composed of design tools integrated in a CAD framework.

Definition 3.1 A design tool *is a software module for performing a specific design task.*

For electronic design a great variety of design tools exists, and new ones are being developed continuously. We can distinguish between *synthesis tools* and *analysis tools*. Synthesis tools allow the design engineer to enter or (semi-) automatically generate design descriptions. Analysis tools aid the design engineer in the verification of previously synthesized design descriptions. Some tools are to be run in *batch mode*, that is, without any intermediate interaction with the design engineer, others can be operated *interactively*. With the graphics capabilities provided by modern workstations, there is a trend towards more interactiveness at the tool level. For example, batch simulators with textual input and output are being replaced by graphical simulation environments offering facilities for stimuli editing, interactive simulation control and waveform display.

We define the term *tool integration* as follows:

Definition 3.2 Tool integration *is the activity of incorporating a design tool into a design environment in such a way that it can be operated as a consistent part of this environment.*

There are different aspects to tool integration, such as the use of a common design database, the use of common design representation formats (tool interoperability), conformance of tool invocation procedures, and the integration of the user interfaces of design tools. A truly *open* and *efficient* CAD framework must allow different kinds of design tools to be integrated conveniently and operated effectively in a single environment.

3.2 GLOBAL RUN-TIME VIEW

3.2.1 The Kernel-Tool Model

As a first step towards a global framework model, we position the design tools with respect to the framework. The framework is to provide a common operating environment for design tools and is the basis for tool integration. In an integrated design environment the design data are unified in the design database. Design data logically reside in a single place and access to design data is controlled by a single authority, the framework. The framework employs uniform concepts for the organization of the design data. Individual pieces of design data may be *shared* among several different design tools, in the sense that each of those tools may have access to the same piece of data. Such sharing may even be performed concurrently. In order to obtain access to design data, the design tools have to interact with the framework.

These observations are reflected by the run-time view of a framework based design environment presented in Figure 3.1. Multiple design tools, being operated concurrently by end-users at different locations in the distributed hardware environment, interact with the framework to obtain access to design data to perform specific design tasks. The framework operates as design database, and performs data and design management functions as the interaction with the tools takes place.

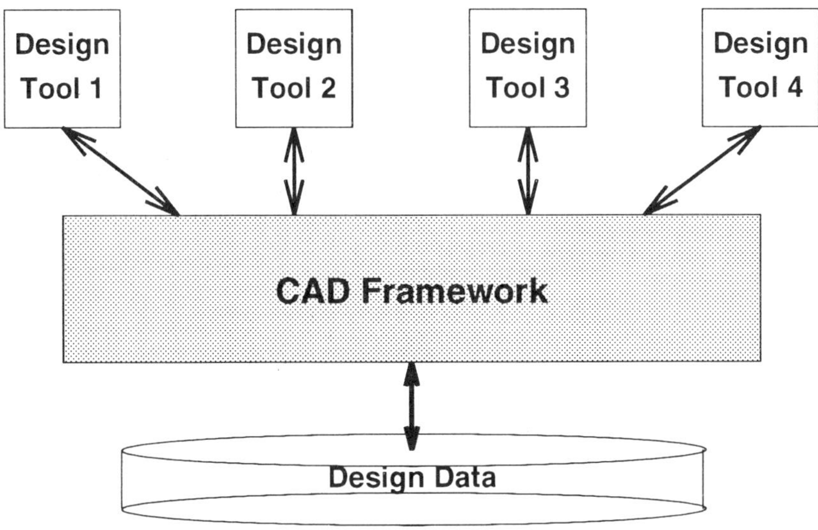

Figure 3.1　Design tools interacting with the CAD framework.

Note that we view the issue of integration from a data perspective. This brings us to the 'database architecture' of Figure 3.1, consisting of services on top of a common repository. Another aspect often considered in the context of frameworks, is the integration of the user interfaces of design tools [CFI92a]. In our view this type of integration is a matter of co-existence rather than a matter of actual co-operation among tools. Integration of user interfaces can be achieved through compatibility of the graphics facilities employed by the tools and through conformance of the look & feel of the user interfaces. Taking into account the emergence of the X Window System and OSF/Motif, we will not consider the integration of user interfaces in detail.

The framework supports the end-user in organizing his design information and managing the design process. This brings up the need for the end-user to interact with the framework, to get informed about the status of his design or to initiate some framework action. For example, he may want to be informed about the version history of a design and subsequently want to select one of the versions for further use. The requirements for this type of interaction are in many respects the same as for the execution of design tasks with design tools: end-users must be allowed to interact concurrently with the framework from different locations in the distributed hardware environment. We therefore broaden the notion of tool, to include design tools as well as a new class of tools, the *framework tools*.

Definition 3.3 *A* framework tool *is a domain neutral tool aimed specifically at interaction of the end-user with the framework.*

Framework tools do *not* access design descriptions to perform specific design tasks. They are *domain neutral*, that is, independent of a specific application domain.

We thus follow the CFI point of view that all interaction with the end-user takes place through tools, which may be framework tools. The CAD framework itself now consists of a *framework kernel* and *framework tools*. The framework kernel is passive and waits for control to be passed from a tool.

Each tool has its own internal operation and, while running, maintains an internal state. The framework kernel does not have to be aware of all details of the internal operation of the tools, but the procedure for the interaction between tools and framework kernel has to be such that the kernel can at all times respond correctly to requests performed by tools. In chapter 6 we will discuss in more detail the interaction between tools and framework kernel, and the possible integration techniques to establish this interaction. In chapter 7 we will discuss the operation of tools and framework in a distributed computing environment.

With this kernel-tool model we basically follow the model of the Unix operating system. In Unix, users do not interact directly with the Unix kernel itself. Instead, a rich set of utilities, including command shells, allows them to interact indirectly. This provides flexibility as users can decide which utilities to use when, where, how long, etc. Also users can add new utilities via the appropriate interfaces.

3.2.2 The Framework Kernel as Transaction Processing System

A dominant aspect of a CAD framework architecture is the *concurrency control* strategy. As is also illustrated by Figure 3.1, multiple tools issue requests concurrently, typically without any mutual coordination, to operate on the data held by the framework. The framework kernel has to handle these requests in such a way that the *state of design* is maintained correctly.

Also in case of system or program failures, the framework is responsible for correctly preserving the state of design. The latter is typically referred to as *recovery*.

We characterize the framework kernel as a *transaction processing system*, which performs transactions on the design data to take it from one consistent state to the next consistent state.

Definition 3.4 *A* transaction *is a sequence of operations that is either performed completely or not at all. It is a logical unit of work, which transforms a consistent state of the database into another consistent state.*

The transaction concept provides convenient means to solve the problems of concurrency and recovery [Dat86]. A transaction has the following three properties [Mul89]:

1. *Failure atomicity*:   Either all operations are performed successfully or their results are undone when a failure occurs.

2. *Permanence*:   The results of committed transactions will not be lost.

3. *Serializability*:   The results of executing transactions concurrently are the same as if they were executed serially.

These properties make a transaction the unit of recovery and concurrency.

Two special operations are the key to providing atomicity of transactions: commit and rollback [Dat86]. Executing either a commit or a rollback operation, after having performed some database operations, establishes a *synchronization point*. A synchronization point corresponds to the end of a logical unit of work, and hence to a point at which the database is in a consistent state. The commit operation signals *successful* end-of-transaction: all updates performed by the transaction are "committed" or made permanent. The rollback operation signals *unsuccessful* end-of-transaction: all updates performed by the transaction are "rolled back" or undone.

To ensure that concurrent transactions do not interfere with each other's operation, i.e. to provide serializability, a transaction processing system employs some kind of concurrency control mechanism. One such mechanism is known as *locking*[1]. The principle of locking is that a transaction can acquire a lock on an object to prevent other transactions from performing conflicting accesses on that object. When a transaction requests a lock on an object, this is granted only if there is no conflicting lock on that object. A locking mechanism typically employs different *kinds* of locks. A *compatibility matrix* [Dat86] specifies for which combinations conflicts occur. An example compatibility matrix is presented in Figure 3.2. For the two kinds of locks, *Exclusive* and *Shared*, the matrix indicates compatibility of Shared locks and conflicts for all other combinations.

	Exclusive	Shared
Exclusive	No	No
Shared	No	Yes

Figure 3.2   Lock type compatibility matrix.

A locking technique which is proven to be correct in general cases is the *two-phase locking mechanism* [EGLT76]. In a two-phase locking system, no locks are released until all locks for the transaction have been acquired. Once a transaction releases a lock, it may not acquire any other lock. The disjointedness of the acquisition and release phases guarantees the correctness of the mechanism. Locking can introduce the problem of *deadlock*. Deadlock is a situation in which two or more transactions are in a simultaneous wait state, each one waiting for one of the others to release a lock before it can proceed [Dat86].

As a result of requests issued by tools, transactions get executed by the framework kernel on the state of design. Specification of the possible requests and the corresponding behavior in terms of transactions is a crucial aspect of a framework architecture.

Transactions performed in the design environment may be of *long duration* [Buc84, BKK85, Kat85a, WvLvdW88]. According to our performance requirement, serving a request for one tool may not cause significant delays in completing requests issued by other tools. This implies that classical techniques, which may e.g. force transactions to wait, are not appropriate and that a high degree of concurrency in executing transactions is to be provided.

---

[1] Locking is not the only possible approach to the concurrency control problem, but it is the one most commonly encountered in practice.

To provide *atomicity* of transactions, the framework must in all cases be able to undo intermediate results of a running transaction when a failure occurs. It are these concurrency and recovery requirements that greatly complicate framework construction. From a software engineering point of view, building a framework is very different from building a CAD tool that serially executes a particular algorithm.

For the run-time view presented above, we can draw an analogy with the operation of the control tower at an airport. Multiple airplanes (tools), approaching or leaving the airport, interact concurrently with the control tower (framework kernel) to obtain access to runways and flying routes. The people in the control tower correctly have to maintain the 'state of airport and air space'. They have to operate on this state in such a way that no two airplanes get directed to the same place at the same time. This requires well-defined procedures for the interaction between the airplanes and the control tower, as well as for the internal operation of the control tower. The control tower does not have to be aware of all details of the internal operation of the airplanes, as long as it obtains sufficient information to perform its task correctly.

3.3 COARSE-GRAIN DATA MANAGEMENT

3.3.1 Meta Data and Design Objects

For purposes of *openness* and *flexibility*, the CAD framework is to be a *domain neutral* facility, i.e. independent of a specific application domain. We adopt a clear *separation of responsibilities* between design tools and CAD framework. Design tools are software modules for performing *specific* design tasks. The CAD framework is the *generic* software infrastructure underneath, providing support for arbitrary design tools and design methodologies. It is responsible for providing general facilities for data organization and design process control. Thus, the framework is based on *invariants* that can be recognized in the design environment, rather than the features of a particular tool set or design representation. This is an important contribution to the *openness* of the design environment.

The separation of responsibilities between design tools and CAD framework directly leads to a distinction at the level of the *data* handled in the design environment. The design tools are geared towards handling data specific to the design tasks they are to perform. They consume and produce design descriptions expressed in terms of specific design representation formats. For example, a high level synthesis tool may employ specific design representation formats for representing input specifications and output results. There is no generally agreed upon set of design representation formats encompassing all stages of the design process. New applications employing new design representation formats are still being developed, and always will be developed as long as CAD is alive.

To retain openness and flexibility, the framework should only to a limited extent be involved in handling the design descriptions themselves. Instead, it should focus on maintaining information necessary for providing its data and design management services at a level independent of specific design representation formats. This brings us to the following principle:

Principle 3.1 *The CAD framework maintains information* about *the design data and the design activities, rather than operating at the level of the detailed design descriptions.*

Historically, this information about the design data and the design activities is called *meta design data*, or simply *meta data* [vLvdW85].

Definition 3.5 Meta data *represents information about the design data and the design activities maintained by the CAD framework.*

The meta data represents information such as the presence of design components, relationships between them, and operations that have been performed on them. The framework itself uses the meta data to perform its management tasks, without having detailed knowledge of the actual design descriptions. For this to work, the design descriptions must present themselves as *manageable entities* to the framework. We therefore require the design data to be organized on a per *design object* basis.

Definition 3.6 A design object *is an aggregate of design data that is to be managed by the CAD framework as a single logical unit.*

The framework does not make any assumptions about the actual contents of a design object (e.g. its granularity, or its data formats). A design object may, for example, contain a mask-layout description, a small circuit netlist, a large circuit netlist, or a behavioral description in some high level design language. This *'raw' design data* is operated upon by the design tools. The framework administers meta data about the design objects, such as their name, owner, and representation type. The meta data - design object distinction is key in supporting different design representations for design at multiple levels of abstraction.

The framework manages design data at the coarse-grain level of design objects. We therefore require design objects to show a well-defined domain-independent behavior to the framework, i.e. a set of operations that can be applied to design objects irrespective of the actual data they contain. This must allow the framework to perform its management tasks in which design objects are involved. The definition of the domain neutral interface for design object manipulation by the framework is part of the framework architecture. A definition of this interface will be presented in section 6.6.

Design tools are not permitted to access design objects directly. Access is allowed only after permission has been obtained from the framework: *the framework is in control.* According to our kernel-tool model, a tool has to *interact* with the framework to obtain access rights for a design object when needed, and return access rights when finished. Once access rights have been obtained for a design object, the tool can operate freely on its contents. To allow the framework to keep track of the activities performed by tools, the interaction has to follow a well-defined procedure and the tools have to supply the framework with the relevant information about the activities they perform. Note that a tool may in this case also be a so called *wrapper* which performs the framework interaction for some other *encapsulated* design tool. The definition of the interface via which the tools interact with the framework is a key view of a CAD framework architecture. A definition of this interface will be presented in section 6.8.

As the interaction with the tools takes place, the framework collects its meta data containing valuable information about the structure and status of the design. The framework itself is the principal user of this information, consulting and administering it when deciding on requests performed by tools: the meta data administration is the *framework's notebook* for maintaining the *state of design.* Consistency is a critical issue here, since the meta data administration should at all times correctly reflect the state of the actual design data as contained in the design objects. The meta data - design data consistency is discussed in section 6.4.

The meta data - design object distinction and the corresponding access procedure for design objects shows a strong analogy with the operation of a public library. In a public library the books (design objects) are physically organized in racks. Management is performed at the coarse-grain level of books, not at the level of individual pages or sentences. Using a (computerized) card index, the library personnel (framework) maintains information (meta data) about the books and their availability. All loans, returns and updates of the collection have to go through the library desk to have the administration updated by the personnel. The library personnel does not need to have detailed knowledge of the actual contents of the books in order to provide proper service. After registration at the library desk, a client (tool) can take a book home to read it as he/she likes, till the book is to be returned. The framework-library analogy stops if we consider that in the framework case complex relationships exist between design objects. Another crucial difference is that the design objects are under design, i.e. the 'books' are being 'written' by the clients who borrow them.

The meta data - design object approach is quite generally agreed upon. For example, it is in line with work presented by Katz [KAC86] and Batory [BK85], as well as with the approaches taken by CFI and JCF and the PowerFrame and ValidFrame systems, as described in section 2.2. An example of a system that does not make a clear distinction between meta data and design data contained in design objects is Cosmic [JLL86].

3.3.2 Implications of Coarse-Grain Management

The meta data - design object distinction seems to match naturally with the separation of responsibilities between design tools and CAD framework. But where does it put us in satisfying our requirements? What are the implications of performing management at a coarse-grain level?

Even though the framework does not interpret the design descriptions, it may still fulfil the role of design database, providing a common repository for all design data. Since the framework has no built-in knowledge of the information carried by this data, the design database is to be a *generic* facility. A generic storage facility can handle arbitrary information, provided the information can be structured in terms of the primitives offered by the data model of the storage facility. Such a generic facility thus provides openness to arbitrary design representation formats. An obvious example of a generic storage facility is the Unix file system, which provides persistent storage for byte-stream data in the form of files, without actually interpreting the contents of these files. Another example of a generic storage facility is a database management system (DBMS). For a particular application, a DBMS allows the structure of the information to be *configured*.

We distinguish two levels of tool integration:

1. *Design environment level.* Integration of a tool at the *design environment level* implies that it can be operated from within the framework based design environment. It can be invoked and, when running, it interacts with the framework to have its data managed as design objects. The framework maintains a single administration in which the information about the design objects, i.e. the meta data, is unified.

2. *Design representation level.* Integration at the *design representation level* implies that the tool adheres to the 'standard' design representation formats employed in a particular design environment. This level of integration provides *tool interoperability*, as it allows the tool to communicate and share design data with other tools.

The CAD framework does not *enforce* integration at the design representation level. Tool interoperability primarily is a tool responsibility; real integration can be obtained only if tools are equipped to 'speak the same language'. The framework must at all times allow incorporation of a tool through integration at the design environment level only, that is, while retaining its own non-compatible design representation formats. CFI is addressing standardization at both levels of integration. An example of standardization at the design representation level is the CFI DR-PI (Design Representation Programming Interface: Electrical Connectivity) [CFI92b]. In chapter 6 we will further discuss the design database. We will then also show how *optional* support for specific design representation formats fits in the framework architecture.

The individual data and design management services, such as version management and access control, operate on meta data, referring to design objects as logical units. This helps them to be *domain neutral* services, independent of the actual data stored in these design objects by the applications. For some services, such as consistency management, the coarse-grain operation may be felt to be somewhat restrictive. Obviously, the implied generic nature requires us to sacrifice application-specific support at the level of the detailed design descriptions. The challenge we face is to match *effectiveness* with the domain neutral character of the framework services. An important aspect in this is the *configurability* of the services, allowing developers and end-users to tune framework behavior to the needs of their application.

It is our ambition to make the CAD framework the electronic assistant of the designer. An important step in this direction is to make the information contained in the meta data available to the end-user. For this purpose special framework tools, called *framework browsers*, enable the end-user to interact with the framework to get informed about the structure and status of his design. This will help him to keep track of his design so that he can quickly decide which tool to apply on which part of his design. In chapter 6 we will present some framework browsers, illustrating that a CAD framework can indeed become the electronic assistant of the designer.

The meta data can be further exploited by putting it at the disposal of design tools. These may fruitfully use information about the verification statuses of design objects or relationships between them. An example is a verification tool that incrementally operates on hierarchical designs to selectively process only those design objects that have been changed since the last run.

The coarse-grain operation of the framework is key in achieving *run-time efficiency*. The framework can focus on efficient handling of the relatively small amount of meta data, tuning to the characteristics of meta data access. Since framework intervention in handling detailed design data is minimized, design tools are not hampered in efficiently accessing their design descriptions.

3.4 DESIGN TRANSACTION MODEL

In section 3.2 we recognized the need for a well-defined procedure for the interaction between tools and framework kernel. In section 3.3 we saw that one important aspect of this interaction is the access of design tools to design objects. We now study this type of access in more detail.

In section 2.1 we reported the failure of the application of conventional DBMSs for the purpose of design database. This failure was due to critical differences in characteristics of data and data access of business applications on the one hand and design applications on the other hand. Design tools typically request all the information pertaining to a piece of design to operate on it for a possibly long period of time [Sid80, Kat82, Buc84]. This points to two important characteristics. First, design tools deal with *collections of related data* which are manipulated as a *single entity*. The retrieval of a single transistor rather than a complete circuit netlist is seldom very meaningful. Second, design tools perform possibly complex operations that may *take a long time* (possibly hours) to produce valuable results. Compare the complexity and duration of a check-and-update operation of an employee's salary in the salary book-keeping of a commercial enterprise with the execution of a design-rule-check on the mask-layout of a new million-gate microprocessor design. The *nature of the data* involved and the *long duration* of design tool operations call for a special treatment of the access of design tools to design data.

The concept of *design object*, as introduced in the previous section, allows tools to store and access related data as a single entity. We introduce the notion of *design transaction* to control the access of design tools to design objects:

Definition 3.7 A design transaction *is a transaction performed by a design tool on a design object.*

A design transaction is a *transaction* in the sense of Definition 3.4. It is an atomic unit of work that takes a design object from one consistent state to another consistent state. A design transaction has the following properties:

- *A design transaction involves a possibly large amount of data.*

  The unit of access involved in a design transaction is a design object which, as explained above, has an undefined granularity and hence may contain a large amount of data. This is in contrast to transactions typically found in business applications which touch small amounts of data.

- *A design transaction may be of long duration.*

  A design tool may operate on the contents of a design object for a possibly long period of time before the object has reached a new consistent state. This is in contrast to transactions typically found in business applications which are of short duration.

- *A design transaction is the unit of concurrency.*

  Concurrent design transactions do not interfere with each other's operation.

- *A design transaction is the unit of recovery.*

  Upon failure, the design object in question is to be returned to its last consistent state.

We adopt the following model for design transactions, which is based on the notion of *conversational transaction* described by Lorie [LP83].

A design transaction is initiated by a CheckOut request issued by a tool. The framework checks availability of the design object in question and, if permitted, makes the object available to the requesting tool. This implies putting a copy[2] of the design object at the disposal of the tool and handing over an *access key* to the tool for operating on the design object. The framework administers a *non-volatile lock* to register that the design object has been made available to the particular tool. The tool may now operate on the design object for an undefined period of time.

A design transaction is terminated by a CheckIn request, with either mode commit or mode rollback. If the tool requests a commit, it has produced valid results and the updated design object is to be returned to the shared repository. If the tool requests a rollback, the updated design object is discarded and the design object remains in its original state. In both cases the corresponding lock is removed.

As is illustrated in Figure 3.3, a design transaction corresponds to the period of time from the CheckOut to the corresponding CheckIn. A single tool is allowed to execute multiple design transactions in parallel on multiple design objects. Concurrent execution of non-conflicting design transactions is crucial for effective multi-user support. Multiple design tools must be allowed to operate concurrently on (possibly non-disjunct) sets of design objects, provided no conflicts occur. The framework is aware of all running design transactions, as they have been registered in its lock administration. Upon a CheckOut request, the framework decides whether the requesting design tool is allowed to execute the transaction on the requested design object, concurrent to the design transactions that are already in progress on the same or other design objects. An extended lock type compatibility matrix (see section 3.2) may be employed to decide on concurrency conflicts. Special lock kinds may be introduced to permit run-time sharing of design objects between collaborating design tools.

---

[2]This does not necessarily imply that all data contained in the design object is to be physically copied. The term copy is merely used to reflect that the original design object is retained while the design transaction is in progress.

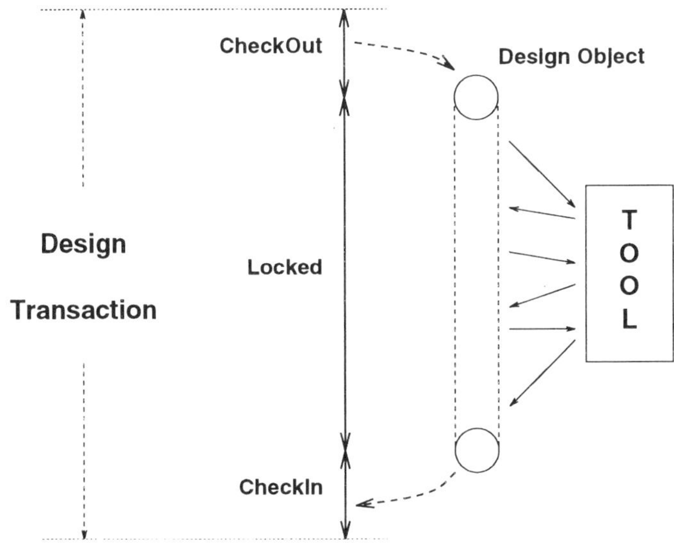

Figure 3.3   A design transaction is initiated by a CheckOut and terminated by a CheckIn. In between CheckOut and CheckIn the tool operates on the design object, which is locked to control concurrent accesses.

Since design transactions may be of long duration, they may prohibit access for conflicting design transactions for a long period of time. In business applications, requesting transactions typically get suspended for a short period of time upon concurrency conflicts. This tactic is not acceptable in design applications. Upon conflict, access to a design object is to be refused and control is to be returned to the requesting design tool. The design tool, on its turn, may return control to the end-user who may proceed with other, non-conflicting design activities. Since transactions do not get into a wait state, deadlock will not occur.

Since the locks on design objects are non-volatile they will survive system crashes and the framework will at all times know which design transactions were in progress on which design objects. This enables the framework to perform *recovery* upon a tool crash by rolling back the design transactions that were in progress for that tool.

As described above, a design tool may return an updated design object upon CheckIn. On this occasion the framework can decide to perform *versioning* by storing the updated design object as a new version while retaining the original design object.

The framework does not interpret the contents of design objects. Hence, it is the design tool's responsibility to deliver a self-consistent design object. The framework has to provide services for consistency among design objects.

The term *design transaction* is widely used but, unfortunately, its definition has not been standardized. Randy Katz uses the term design transaction to denote a sequence of database operations that map a consistent version of a design into a new consistent version [Kat85a]. The term 'consistent' refers to application-level consistency, i.e. correctness of the design as demonstrated by analysis tools. A design transaction may include several tool runs for the creation and verification of the design. Katz states that design transactions are *non-atomic* units of design consistency. Intermediate states may become visible to concurrent transactions and recovery mechanisms may restore the database to an intermediate state.

The Damokles system [RRR$^+$88] uses the term design transaction to denote the manipulation of multiple design objects in a private area. A design transaction may comprise multiple tool runs. A CheckOut and CheckIn operation are used to transfer design objects to/from the private area from/to a public database.

In contrast to Katz and Damokles, but in line with e.g. Lorie [LP83], we have chosen a design transaction to be an *atomic* unit of work in the classical sense. It is performed by a single design tool on a single design object, and hence is of finer granularity than the design transaction of Katz and Damokles. It is the basic unit of consistency for operation by design tools on design data. It provides the appropriate primitive for the framework to fulfil its role of design database, i.e. to arbitrate among multiple concurrently running tools requesting access to design data, and to recover a consistent state upon a failure. Higher level consistency, such as application-level consistency, has to be provided on top of this basic facility.

The transaction model presented above is consistent with the public library metaphor presented in section 3.3. The reader (tool) initiates a loan by requesting a particular book (design object) at the library (CheckOut). He takes the book home for a possibly long period of time, during which it is not available to other readers. The loan is terminated when the book is returned by the reader (CheckIn).

3.5 THE FRAMEWORK ARCHITECTURE DEFINITION PROCESS

We refine Figure 3.1 to reflect conclusions derived in the previous sections. The result is Figure 3.4, representing our *global framework model*. This global framework model is the starting-point for further definition of our framework architecture.

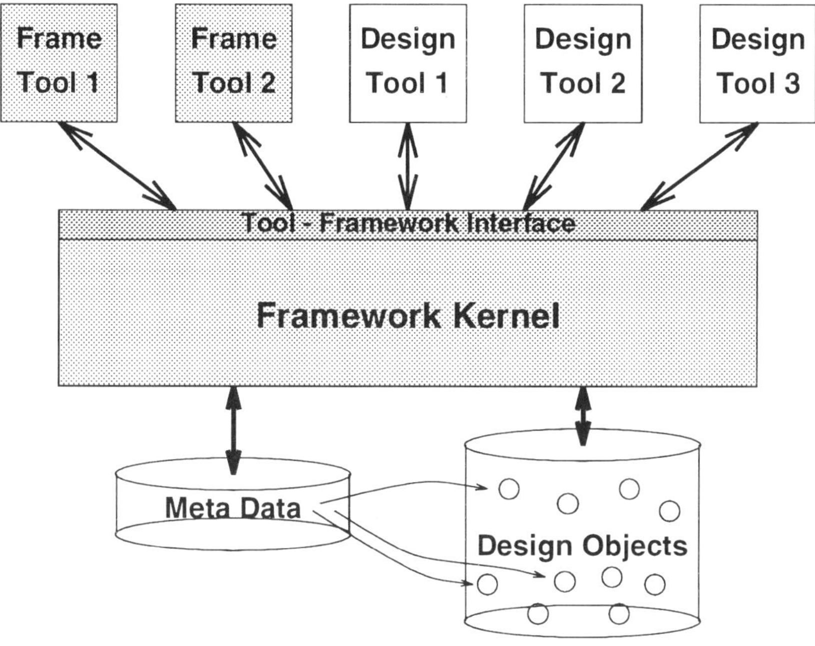

Figure 3.4   Global framework model.

The CAD framework (shaded components) consists of a *framework kernel* and *framework tools*. All interaction with the end-user is through tools. The framework kernel maintains the state of design and controls all access to the data. We logically distinguish between *meta data* and 'raw' design data contained in *design objects*. Meta data refers to design objects as logical units. Interaction between tools and framework kernel takes place according to a well-defined procedure via the *tool - framework interface*. In particular, design tools can perform *design transactions* on design objects, via CheckOut and CheckIn requests.

We must carefully determine which approach to take to the more detailed definition of the CAD framework architecture. This includes selection of proper ways to represent the architecture. In chapter 1 we remarked that ideas on framework architectures do not get communicated well in the EDA community. In the previous chapter, in particular section 2.2, we concluded that a CAD framework is a multi-dimensional software capability that can be viewed from many different perspectives. We have to define which *views* we will use to represent the framework architecture.

It is crucial to realize that requirements on CAD frameworks are still evolving (section 2.3). The detailed functionality of the framework services is subject to the evolution of requirements, and is not an invariant that we consider to be the primary output of the architecture definition process. Moreover, this detailed functionality often is a matter of taste, rather than being a matter of superior versus inferior solutions. We relegate consideration of detailed functionality to a later step in framework development. In its place we promote the activity of modeling the real world. This approach to system development is also advocated by Jackson, who tries to cope with changing functional requirements in [Jac83]. We will start our architecture definition process with the derivation of a model of the design environment. This model provides a vocabulary of well-defined terms, and thereby a context for functional specifications. Obviously, such a model implies function, since it implicitly defines a set of possible functions, but we agree with Jackson that a model is a more stable starting-point than a purely functional specification. Also with respect to user-friendliness, it is good design practice to start by modeling the world our users are to live in, rather than by specifying the system functions.

As our second step we choose to identify the principal framework functions and to define the logical structure of the framework that imposes order on these functions. This logical structure must also indicate where and how to fit in the unspecified detailed functionality of the framework services. To complete the definition of the architecture, we will describe the internals of the framework at the physical level. This will reveal how the framework can actually be realized as a computer-based system.

Via this line of thought we come to the following three views that we will use to represent the framework architecture:

1. *Information architecture*

   The world of our users is the design environment. The *information architecture* describes at a domain neutral level the different types of objects in the design environment and their relationships.

   The CAD framework is the infrastructure for building integrated design environments, which is to provide the concept for logically organizing different types of domain neutral information in the design environment. The information architecture, hence, is a key view on the framework architecture [vdHT90]. Since the information architecture documents the environment that is offered by the framework to its users, rather than revealing aspects of the internal structure of the framework, it is considered a *user-view* of the framework.

   We characterized the framework kernel as a transaction processing system that maintains the state of design. Framework functions interact by operating on a shared set of (meta) data. As is also expressed in [vdH91], the information structure on which the framework operates dominates the algorithmic aspects of its operation. We consider it a matter of good design to start with the specification of this information structure. A first order approximation of the information structure will be given by the definition of the object types and relationships in the information architecture.

   The formal technique of our choice for representing the information architecture is *data modeling*[3]. In the next chapter we will discuss data modeling techniques. These will subsequently be applied in the definition of the information architecture in chapter 5.

2. *Component architecture*

   To provide additional detail on the functionality and logical structure of the framework, we present the *component architecture.* The component architecture identifies the individual framework components and the dependencies between them. It defines the functional decomposition of the framework: principal framework functions are identified and allocated to mutually ordered components. Components make their functions available to other framework components or tools by 'exporting' them via their *interface.* The dependencies between components define which components use functions of which other components (calling dependencies), including constraints on the order of execution of these functions. Since it documents the internal structure of the framework,

---

[3] Data modeling is also referred to as *information modeling* or *conceptual modeling*.

the component architecture is considered a *developer-view* of the framework.

As part of the component architecture we will define some key interfaces. Special attention will be paid to the definition of a well-structured tool - framework interface. The definition of this interface is considered a user-view for tool integrators.

3. *Implementation architecture*

The *implementation architecture* describes the internals of the framework at the physical level. It defines how the framework is implemented in terms of Unix operating system primitives, such as processes, files, IPC mechanisms, etc. Logical framework components get allocated to Unix processes, and physical communication structures get defined. The concurrency aspect plays a key role here. The choices made at this level are key to obtaining maximum efficiency and optimal behavior with respect to physical distribution and multi-user support.

These three views together describe the framework architecture. Starting with a global 'picture' of the design environment in terms of object types and their relationships, they define the principal functions and global structure of the framework as well as the mechanics of its inner structure. We will present the three framework views in the following chapters.

4

DATA MODELING

4.1 INTRODUCTION

We will start the framework architecture definition process with the derivation of a model of the design environment: the information architecture. This
information architecture is to describe the logical organization of the design
environment in terms of object types and their relationships. It will provide
a vocabulary of well-defined terms, and thereby a context for detailed functional specifications. Besides, it will give a first order approximation of the
information structure on which the framework operates.

The information architecture is to be represented in terms of well-defined
primitives. It must be a clear and unambiguous view of our CAD framework architecture that permits framework properties to be communicated and
discussed. We therefore adopt a formal technique for deriving and representing the information architecture: *data modeling*. Data modeling is concerned
with techniques and constructs that support a high-level representation of data
to reflect the real-world situation. Data modeling helps in achieving understanding of the information needs of an organization and by extension the
way in which an organization functions.

During the last two decades there have been many developments in the field
of data modeling. Early approaches in data modeling come from the database area, and were primarily aimed at the structuring of data for purposes
of electronic data processing. Later on, more attention was paid to the formalization of the meaning of the data [Dat86, tB92]. We have also seen the
incorporation of modeling concepts originally devised in the artificial intelligence area [BZ81, BG86]. These developments have led to the availability

of a variety of data modeling techniques.

For our purpose the data modeling technique needs to offer a set of *well-defined modeling constructs*, permitting us to derive a formal representation of the logical organization of the design environment. The data modeling technique may then act as a "language" by means of which information architectures can be discussed. Preferably it also offers a clear and unambiguous *graphical notation*, to further improve communication of information architectures. In this chapter we will introduce the data modeling technique that we will use in the next chapter as a formalized tool for information analysis and representation.

4.2 DATA MODELS

We start from the following definition [Lyn84]:

Definition 4.1 A data model *is a collection of concepts and constructs for expressing the static properties, dynamic properties, and integrity constraints of an application environment.*

A data model is characterized by [TL82, AM84, tB88, tB92]:

- A collection of *constructs* (Data Definition Language, DDL).

- A collection of *fundamental operations* (Data Manipulation Language, DML).

- A set of *integrity constraints* defined on its constructs.

Given a data model, *data schemas*[1] can be defined.

Definition 4.2 A data schema *describes the structure and properties of a specific application environment in terms of the constructs of a corresponding data model.*

---

[1] Data schema is also referred to as *information model* or *conceptual model*.

Thus, a data model can be seen as a generic mechanism out of which data schemas can be instantiated.

Definition 4.3 A database *is a data repository containing a possibly large amount of interrelated data, structured according to a corresponding data schema.*

Thus, a data schema can be seen as a generic description out of which the contents of a database can be instantiated.

Historically, the following four classes of data models can be recognized [Lyn84, tB88, tB92]:

1. *hierarchical data models*

2. *network data models*

3. *relational data models*

4. *semantic data models*

The hierarchical, network, and relational data models are frequently referred to as the *classical data models* [Lyn84, BG86]. The *machine-oriented* hierarchical and network models provide only primitive operations, and the user must deal with aspects of the internal organization. The need for *data independence*, i.e. separation of logical organization and internal organization of the data [Dat86], is generally acknowledged. The relational data model is more user-oriented. Its main attraction is its mathematical clarity, which facilitates the formulation of non-procedural, high-level queries and thus separates the user from the internal organization of the data. However, the relational model too has some serious drawbacks. First of all it is a flat model; the data are to be structured in the form of tables containing atomic values. The use of composed keys does not provide the user with sufficient means to represent all abstractions in a precise way. Integrity constraints have to be defined explicitly; this is not an integral part of the modeling process. In [tB88] Ter Bekke clearly demonstrates various defects of the relational data model.

The classical data models are all *record-based*. When modeling an application environment, not all record types in the resulting data schema correspond to the complete definition of a particular concept from that environment. The classical data models lack semantic expressiveness [AM84, BG86, PM88]. For a more extensive overview of the classical data models and their most important drawbacks we refer to [tB88] and [vdW86].

A relatively recent trend is the incorporation of more semantic modeling capabilities into data models. This has yielded a new class of data models: the *semantic data models*. These data models enable the user to better formalize the semantics of the data, and are therefore considered more *user-oriented*. Instead of being based on the record model, which arranges data in fixed linear sequences of field values, the semantic data models are *object-based*; the application environment is modeled as a collection of interrelated objects, each one corresponding to some concept from this environment.

Quite a number of more or less semantic data models have been defined over the past fifteen years. One of the first useful proposals in this area was the *entity-relationship model* [Che76]. Attempts to categorize semantic data models are described in [Lyn84], [AM84], and [PT88]. In [PM88] Peckham et. al. describe the generic properties of semantic data models and present a selection of models that have been proposed since the mid-1970s. In [vdW86] we concluded that the OTO-D[2] semantic data model [tB88, tB92] is an attractive data model for defining and representing the information architecture of a CAD framework. "OTO-D" stands for Object Type Oriented Data model. It offers well-defined modeling constructs, an attractive query language and has a clear way of visualizing the data schemas one defines. For a good understanding of the data schemas that are derived in the next chapter, we first present the OTO-D data model. We refer to [tB88] and [tB92] for a more extensive description.

---

[2]The OTO-D data model has been developed at the Mathematics & Computer Science department of the Delft University of Technology, under the direction of dr. J.H. ter Bekke.

4.3   THE OTO-D SEMANTIC DATA MODEL

4.3.1   Data Definition

Types

In the semantic approach the notion of *abstraction*, i.e. representing relevant details while suppressing irrelevant ones, plays a dominant role. When representing the invariants of a dynamic environment, not the elements themselves but their properties are important. The OTO-D approach is therefore based on the notion of *type*.

Definition 4.4 *A* type *is a definite aggregation of distinct properties.*

For example the abstraction:

```
TYPE Student = Name, Address, Department
```

defines a student as being completely characterized by the properties Name, Address and Department. These properties are called *attributes*. An object having properties of a certain type is called an *instance* of this type.

A type definition is a positive statement or *assertion* about the application environment. It consists of two parts: a *subject* and a *predicate*. The subject denotes the new concept and the predicate denotes the collection of known properties by means of which the concept is described. A data schema consists of a number of these type definitions.

Convertibility and Relatability

OTO-D offers two concepts to discuss relationships within and among type definitions, respectively. The first concept, named *convertibility*, is related to the internal structure of an assertion. Because the subject (type) is completely characterized by the predicate (attributes), while on the other hand each predicate describes one subject, there is a one-to-one correspondence between the subject and the predicate of an assertion. Convertibility also has consequences at the instance level. Each object is uniquely characterized by its attribute values; the instance identification is of no importance. Based on the notion of convertibility, the type definitions can be checked for completeness during the construction of the data schema.

The second concept, *relatability*, applies to the relationships among the different assertions of a data schema. It means that an attribute is related to the definition of the type with the same name. For example:

```
TYPE Department = Dept_Name, Head
TYPE Student = Name, Address, Department
```

The attribute Department in the definition of Student is related to the definition of type Department. Thus, object types are defined in terms of previously defined object types, which may be base types[3]. Relatability also has consequences at the instance level. It implies that an attribute value is related to one and only one instance of the type to which the attribute refers. Relatability thereby automatically fixes *referential integrity constraints* at the instance level. For example, consider the attribute relationship between Student and Department in the example above. As a consequence of relatability the set {Student ITS Department} is at any time a subset of {Department}: a property referred to as *subset invariance*.

Aggregation and Generalization

OTO-D offers the abstraction primitives *aggregation* and *generalization* to construct a data schema. This indicates that OTO-D has roots in the work of D. Smith and J. Smith, who originally introduced these two modeling concepts [SS77]. *Aggregation* is a form of abstraction in which a number of different properties is combined to create a new named object, about which we can talk without having to bother about the underlying objects. Examples of aggregations were the definitions of the types Student and Department. Together they constitute an *aggregation hierarchy*.

OTO-D offers a clear diagrammatic notation to visualize the relationships among the types defined in a data schema. The graphical representation of the example data schema defined above is depicted in Figure 4.1. Each attribute relationship is represented by an edge connecting the bottom of the compound type to the top of the attribute type. The referential integrity constraints along these edges are always satisfied: an instance of a compound type is existence-dependent on the instances of its attribute types.

It may happen that a type appears more than once as an attribute in a type definition, to fulfil different roles in the definition of the compound type.

---

[3]For reasons of conciseness we will omit the explicit definition of base types from the textual definitions of our example data schemas.

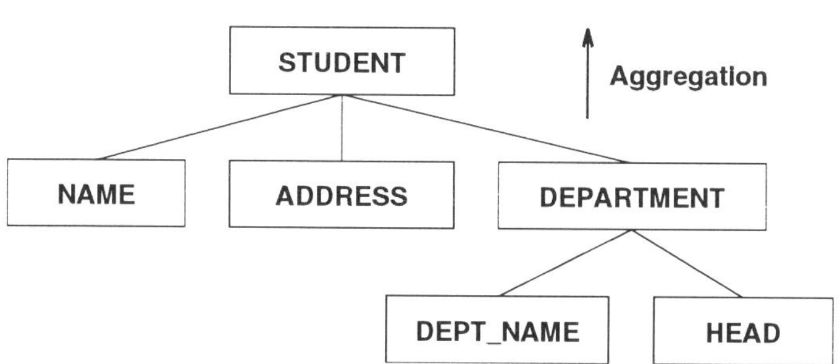

Figure 4.1   Example of an aggregation hierarchy.

For this purpose OTO-D offers *role attributes*. Role attributes are denoted by a prefix followed by the name of the type to which the attribute refers, separated by a dash '-'. For example, if for each student a second (family) address is to be administered where he can be reached during the holidays, the type Student can be defined as follows:

```
TYPE Student = Name, Home-Address, Holiday-Address,
                                            Department
```

In the diagrams the prefixes of role attributes appear as labels of the attribute relationships.

The second essential modeling construct of OTO-D is *generalization*. This construct is not present in the classical data models. Generalization is a form of abstraction that relates one or more type(s) to a more generic type. For example, Person is a generalization of Student. This form of abstraction is known in knowledge representation research in the artificial intelligence area as the IS-A relationship: student IS-A person. Generalization helps in identifying relevant new object types by considering common properties of other types. The opposite of generalization is *specialization*. It is used to define a specific type given a more general one. A number of types that are related this way, together constitute a *generalization hierarchy*. In the following example the type Student is a specialization of the type Person. This is indicated by the square brackets around Person in the definition of Student.

```
TYPE Department = Dept_Name, Head
TYPE Person = Name, Address
TYPE Student = [Person], Department
```

The graphical representation of this data schema is depicted in Figure 4.2.

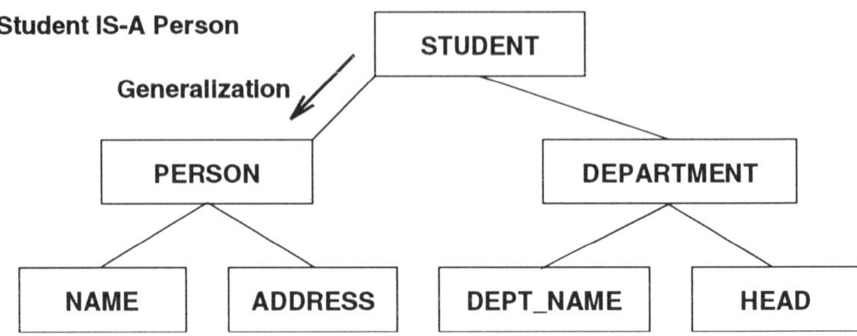

Figure 4.2   Example of the use of generalization.

Notice that in the diagram the relationship between the object types that constitute a generalization hierarchy is depicted by means of an edge connecting the corners of the respective rectangles.

Generalization allows concepts of an application environment to be described at different levels of abstraction. For example, at a global level a person is described by the properties Name and Address, while at a more detailed level the persons that are student are described by the properties Name and Address (because a student is-a person) supplemented with Department. Notice the so called *attribute inheritance*. As a result of the convertibility, OTO-D rejects specializations that do not contain additional attributes. Generalization allows objects to belong to more than one type. At these different types the object can be known under different names: *synonyms*. Also notice that as a consequence of the subset invariance (relatability) at any time the set of students is a subset of the set of persons.

Cardinalities

While defining object types and attribute relationships among object types with OTO-D, the cardinalities of relationships get defined as well. This is not a separate activity and requires no additional constructs. Each instance of a compound type has a single value for each of its attributes, but multiple instances of a compound type may have the same value for a particular attribute. For example, according to the data schema of Figure 4.1, each student is related to a single department, but multiple students may be related to the same department. Figure 4.3 presents a short reference for reading cardinalities from a data schema.

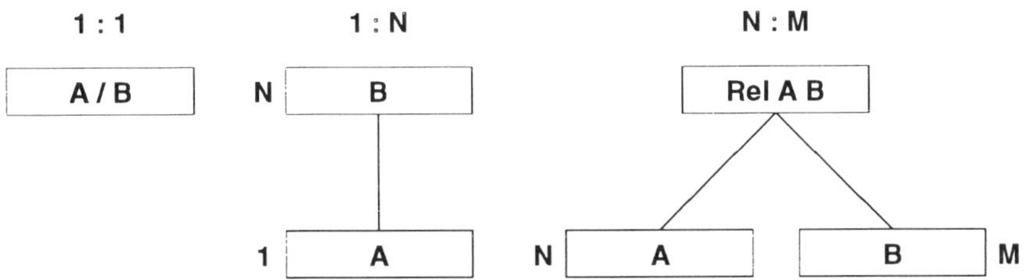

Figure 4.3 Short reference for reading cardinalities from a data schema.

The configuration on the right-hand side in Figure 4.3 corresponds to an N:M relationship, or many-to-many relationship, between types A and B. The type RelAB relates instances of types A and B on a pairwise basis. Each instance of type A may thus be related to multiple instances of type B and each instance of type B may be related to multiple instances of type A.

A Modeling Example

To further illustrate the OTO-D modeling capabilities, we present an additional example. We extend the data schema of Figure 4.1, to also model courses and the enrollments of students for courses. A course has a name and a detailed description and is offered by one of the departments of the university. A student may enroll for multiple courses and multiple students may enroll for the same course. For each enrollment of a student for a course we have the result the student may obtain. This brings us to the following data schema:

```
TYPE Department = Dept_Name, Head
TYPE Student = Name, Address, Department
TYPE Course = Course_Name, Info, Department
TYPE Enrollment = Student, Course, Result
```

The graphical representation of this data schema is depicted in Figure 4.4.

The object type Course aggregates the relevant properties of a course. The object type Enrollment models the many-to-many relationship between students and courses, with a result for each enrollment. Possible values for the Result attribute are 'not yet completed', 'dropped out' or the grade obtained upon completion.

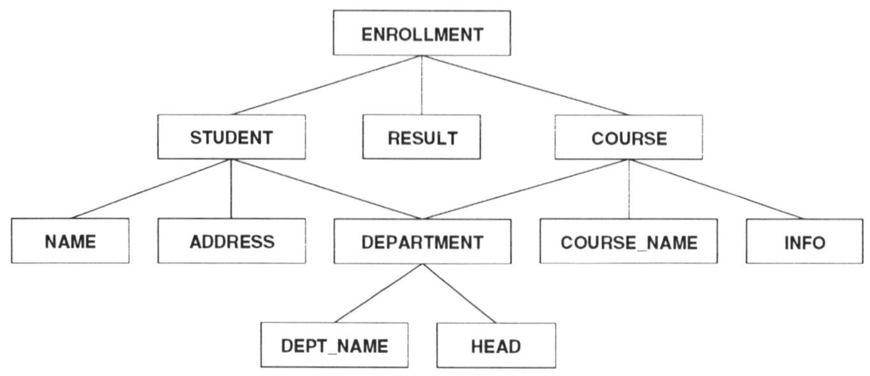

Figure 4.4 Data schema for the student - course example.

4.3.2 Data Manipulation

We present a concise overview of the Data Manipulation Language (DML), or query language, of OTO-D. This DML will be used in the next chapter to demonstrate that relevant information can be retrieved on the basis of the data schemas that are defined to model the design environment.

The Data Manipulation Language of OTO-D is based on the semantic concepts of the data model (convertibility and relatability). It offers *selection*, *extension* and *modification* commands. An important expression is the selection expression, the general form of which is:

```
<type name>            ⇐    subject type
ITS <attributes>       ⇐    property list
WHERE <condition>      ⇐    qualifying predicate
```

Starting from a subject type (an arbitrary compound type which is chosen to be the subject of the query), attribute types can be addressed via the ITS construct. The ITS construct permits downward traversal along the attribute relationships defined in the data schema. As a consequence we can only "look downward" along the schema, starting from the subject type to ITS attributes, ITS attributes ITS attributes, etc. The semantic concepts of OTO-D guarantee that all data that can be addressed are present (referential integrity) and related in a meaningful way according to the data schema. The property list in the selection expression specifies the attributes for which values have to be retrieved. The qualifying predicate specifies value restrictions on attributes, in order to select the instances of the subject type that must be taken into account.

Using the keyword GET, selection commands can be formulated. Consider, for example, the data schema of Figure 4.4. If we want to retrieve the names and addresses of all students at the Anthropology department, this can be done by issuing the following query:

```
GET Student
   ITS Name, Address
   WHERE Department ITS Dept_Name = 'Anthropology'
```

If we want to retrieve information on all courses that a student named Latimer from the Anthropology department has enrolled for, this can be done as follows:

```
GET Enrollment
   ITS Course ITS Course_Name,
       Course ITS Department ITS Dept_Name,
       Result
   WHERE Student ITS Name = 'Latimer'
   AND Student ITS Department ITS Dept_Name =
                                      'Anthropology'
```

Set functions, such as COUNT or MAX, are available to state more complex requests.

Extension commands can be used to derive information that is not directly available from the data stored in the database. The general form of the extension command is:

```
EXTEND <type name>
   WITH <attribute name> = <extension definition>
```

It attaches a temporary attribute to a type according to an extension definition. Subsequent selection commands can use the attribute as if it were a permanent one. An example extension command for the data schema defined above (see Figure 4.4) is:

```
EXTEND Department
   WITH NumberOfStudents = COUNT Student PER Department
```

To modify the contents of the database, three types of modification commands are available: INSERT, DELETE and UPDATE.

4.3.3   Constraint Definition

Upon definition of a data schema with OTO-D, several integrity constraints get defined implicitly. These constraints directly follow from the application of the constructs offered by the Data Definition Language, and are referred to as *inherent constraints*.

In addition to the defined structure and its inherent constraints, OTO-D allows extra constraints to be specified explicitly: *explicit constraints*. One class of explicit constraints are the *static constraints*. A static constraint is a statement which is true for every valid database state with a defined structure. Static constraints can be defined according to the following syntax:

```
ASSERT <type name>
  ITS <virtual attribute> <representation>
                            = <extension definition>
```

An example static constraint for the data schema defined above (see Figure 4.4) is:

```
ASSERT Department
  ITS NumberOfStudents (1..*)
                        = COUNT Student PER Department
```

In this example the representation of the virtual attribute NumberOfStudents is the value range (1..*). It specifies a condition for the number of students related to a department: at least one. Another constraint that one may like to have enforced is that students can enroll only for courses offered by their own department. This is expressed as follows:

```
ASSERT Enrollment ITS ValidDept (TRUE) =
        Course ITS Department = Student ITS Department
```

4.4 DISCUSSION

Starting from the semantic concepts of convertibility and relatability OTO-D provides a sound data model to construct data schemas using the abstraction primitives aggregation and generalization. Summarizing, the following are considered the strong points of the OTO-D data model:

- OTO-D is *object-based*: the design environment can be modeled as a collection of interrelated objects, each one corresponding to some concept from this environment.

- The Data Definition Language (DDL) has a high level of semantic expressiveness. It offers a *small number* of *well-defined* modeling constructs. The definition of integrity constraints is an integral part of the modeling process.

- OTO-D offers an *attractive graphical notation* to represent the object types and their relationships defined in a data schema. This is a strong point for documenting and communicating information architectures. Moreover, it permits the construction of an attractive graphical query interface for meta data access.

- The Data Manipulation Language (DML) is comprehensible and expressive. It matches perfectly with the concepts that are used at data definition time.

In brief, OTO-D is simple, well-defined, expressive, and user-oriented. We conclude that OTO-D provides an appropriate formal means to represent concepts from the design environment in a clear and unambiguous way as object types and their relationships.

A drawback of OTO-D is that it is not very well-known, at least when compared to various other data models such as the entity-relationship model (see below). It has also been mentioned that modeling with OTO-D is not straightforward. However, we consider it an advantage that OTO-D imposes some modeling discipline as this promotes careful definition of the precise structural semantics of application environments.

As we mentioned in section 4.2, a variety of more or less semantic data models has been defined over the past years. We will not outline variant data modeling techniques in detail here, but would like to make a few remarks on some approaches.

One of the first steps towards developing a higher level data model was the *entity-relationship model* (E-R model), proposed by Chen [Che76]. With the E-R model, a data schema is defined in terms of *entities* and *relationships* among entities. Both entities and relationships can have attributes. Data schemas are represented pictorially by so called E-R diagrams. The E-R model has an increased orientation towards the user and is a convenient database design tool. Unfortunately, the E-R model makes a strict distinction between entity and relationship. It does not allow any specific real-world concept to be both an entity and a relationship. Consider, for example, the concept of "marriage", which can be seen as a relationship between two persons or as a legal entity. The OTO-D data model, on the other hand, has one fundamental structuring concept, allowing real-world concepts to be modeled as both an entity and a relationship in one and the same data schema. Such flexibility of interpretation is fundamental to the semantic approach. The graphical notation of OTO-D is simple and yields ordered diagrams in which no additional labeling is required to represent cardinalities, as must be done in the E-R diagrams.

Another category of semantic data models are the *functional data models*, a representative of which is Daplex [Shi81]. A functional data model views objects, attributes of objects and inter-object relationships uniformly, and defines them as functions. For example, zero-argument functions define object types, and single-argument and multi-argument functions define relationships among object types. The Daplex data model has succeeded in combining functional modeling with the abstraction primitives aggregation and generalization. It offers a uniform query facility based on functional composition. However, we do not consider Daplex as a superior modeling technique for defining our information architecture. The use of multi-argument functions tends to make relevant object types implicit. A key drawback is that the definition of the structural semantics in terms of object types and (functional) relationships among object types does not inherently specify all retrieval possibilities. The reason is that the functional relationships can be traversed in one direction only. Additional derived functions must be defined to allow retrievals in reverse directions. The OTO-D data model, on the other hand, allows the structural semantics expressed in a data schema to be used in both directions upon data retrieval. Consider for example Figure 4.1. Given a student, information about its department can be retrieved:

```
GET Student
   ITS Department ITS Dept_Name
   WHERE Name = 'Hubbard'
```

In a similar way information about students can be retrieved for a particular department:

```
GET Student
   ITS Name, Address
   WHERE Department ITS Dept_Name = 'Anthropology'
```

No additional definitions or constructs are required.

CFI employs a data modeling technique in its standardization activities known as Express [Exp91]. Express is a language that permits entities to be defined in terms of attributes, constraints, and operations on those entities. Express also has an associated graphical notation for the display of data schemas (or information models), dubbed Express-G. This notation supports only a subset of the Express language. In several respects Express compares to OTO-D. An attribute of an entity relates to either a base type or another entity. Express does not distinguish between attributes and relationships explicitly. It offers Subtype - Supertype constructs, equivalent to specialization - generalization, and supports multiple inheritance. A drawback of Express is that a relationship between two entities typically gets defined in two places, i.e. in both entity definitions. For example, according to the base connectivity model of CFI [CFI92b], a Library object groups together a collection of Cell objects. This is defined in Express as follows:

```
ENTITY cfidrLib
   ...
   Cells:    SET [0:?] OF cfidrCell;
   ...
END_ENTITY;
ENTITY cfidrCell
   ...
   Owner:    cfidrLib;
   ...
END_ENTITY;
```

The entity cfidrLib has the attribute Cells, while the entity cfidrCell has the attribute Owner to reference its library. This compares to the reported drawback of the Daplex functional data model in that retrieval possibilities have to be specified explicitly. The Express example is defined in OTO-D as follows:

```
TYPE Library = LibName, ...
TYPE Cell = Library, CellName, ...
```

That is, the relationship between Cell and Library gets defined in one place, via a single attribute relationship among two object types. As a result of the redundancy in the entity definitions, Express language texts tend to be lengthy and obscure. We think that these problems are caused by a lack of order among the defined entities. This also appears from the Express-G diagrams, which tend to be unordered. The 'double' attributes appear as 'twin' edges in the diagrams. Further, the direction of the edges has to be indicated explicitly, and cardinalities must be annotated.

The above observations confirm our conclusion that the OTO-D data model is a simple, well-defined, expressive, and user-oriented data model, which can be used as a formalized tool for defining and representing the information architecture of a CAD framework.

5

THE INFORMATION ARCHITECTURE

5.1 INTRODUCTION

In this chapter we will apply the OTO-D data modeling technique to define the information architecture of a CAD framework. As stated in section 3.5, this information architecture is to describe at a domain neutral level the logical organization of the design environment in terms of object types and their relationships. In particular we will address the following data and design management topics (see also the framework requirements presented in section 2.3):

- Logical distribution of design data and design activities, including library facilities.

- Design transaction management.

- Version management.

- Support for hierarchical multi-view design.

- Tool management.

- Design flow management.

A data schema will be defined representing the types of information relevant to these topics.

In addition we will show how some other, more peripheral, aspects may be incorporated into the data schema. These schema extensions will not be

71

motivated extensively; choices presented are sometimes a matter of taste, and should merely be taken as examples of how the data modeling technique may be used for representing flavors of these related aspects as well.

The method we employ to derive the data schema contains the components of perception, representation, and validation. We start by looking at the design environment in order to perceive relevant real-world concepts and their dependencies, with special attention for invariant properties. We put on our OTO-D glasses, which give optimal sight to recognize object types and their relationships. New object types are defined in the data schema, and checked for completeness and correctness with respect to other parts of the data schema. The modeling process is *incremental*. We successively focus on different aspects of the design environment to represent them in the data schema. Remember that object types are defined in terms of other (attribute) object types. Attribute types may have been defined at a previous stage and/or may be refined at a later stage. The introduction of new types may cause refinements of previously defined types, for example, to avoid redundancies or to have attributes attached at the proper level in an aggregation hierarchy. Since each variant data schema has a well-defined meaning, alternatives can be discussed on formal grounds.

We want our CAD framework to become the electronic assistant of the design engineer, which provides useful information to help him manage the design process. In the course of defining the data schema we will present example queries to demonstrate that relevant information can be retrieved on the basis of the information structure represented by the data schema.

5.2 DESIGN OBJECTS

In section 3.3 we concluded that the CAD framework manages design data at the coarse-grain level of design objects. Hence, the first object type in the design environment that we may include in our data schema is *DesignObject*. In order to define the object type DesignObject, we have to identify the properties by which a design object is characterized. As an initial step we specify that each design object has a *name* for purposes of identification by the design engineer, a *designer* who owns the design object, and a *date* of construction. In addition each design object has a reference to the physical location where its design data resides. In our public library paradigm (section 3.3) this compares to a reference to the physical location of a book in the racks. The object type DesignObject is defined in OTO-D as follows:

```
TYPE DesignObject = Name, PhysLoc, Designer, Date
```

The corresponding diagram is depicted in Figure 5.1.

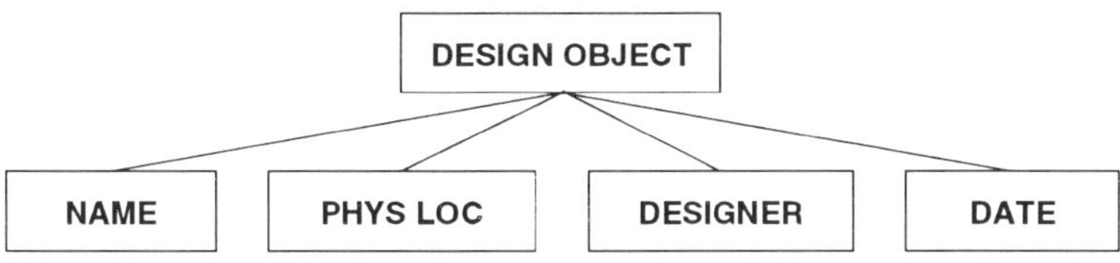

Figure 5.1   The object type DesignObject.

The attribute types of DesignObject can be either base types or compound types. For example, the object type Designer may be defined as a base type. In this case instances of type Designer simply are identifications of design engineers about whom no further information is maintained by the framework. As an alternative, we may define Designer as a compound type, for example:

```
TYPE Designer = Name, Address, Permissions,
                                        ExpertiseLevel
```

This allows the registration and use of more detailed information about individual design engineers.

5.3   PROJECTS

The initial data schema defined in the previous section does not yet include a major principle for the organization of design objects in the design environment: Apparently there is one big pool of attributed design objects. We introduce the notion of *project* to provide *logical distribution* of design data and design activities.

Definition 5.1 A project *is a local environment in which design activities may be performed.*

A project contains design data, organized as design objects, as well as administrative information. It provides a local context for performing design activities. The design environment is organized as a collection of projects containing the actual design objects. This is illustrated by the intuitive picture in Figure 5.2.

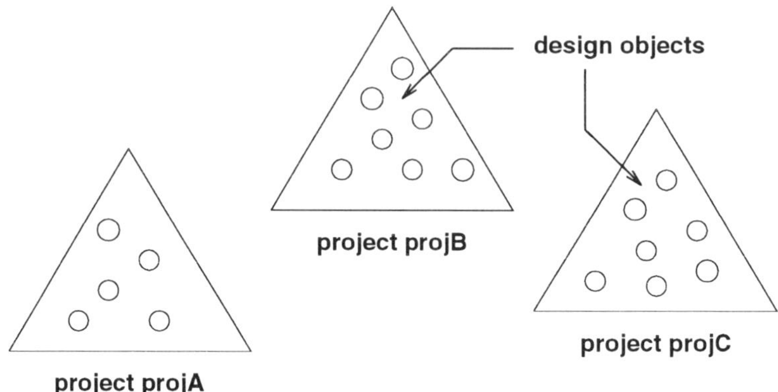

Figure 5.2   The design environment is organized as a collection of projects containing design objects. Triangles denote projects, circles denote design objects.

We have recognized the object type *Project* and include it in our data schema. In the design environment a project is uniquely identified by a *project identification*. A *team* of design engineers is responsible for performing the design activities in a project. Each design object resides in a particular project. Hence, the object type DesignObject obtains an additional attribute Project, to refer to the project the design object resides in. This yields the following data schema (see Figure 5.3):

```
TYPE Project = ProjectID, Team
TYPE DesignObject = Name, Project, PhysLoc, Designer,
                                                    Date
```

We want projects to act as *name spaces* for design objects, that is, the *scope* of a name of a design object is the project it resides in. Design objects have to be named uniquely within their project environment, but design objects from different projects may have identical names. Thus, we require the pair <project, name> to be an unique identification of a design object.

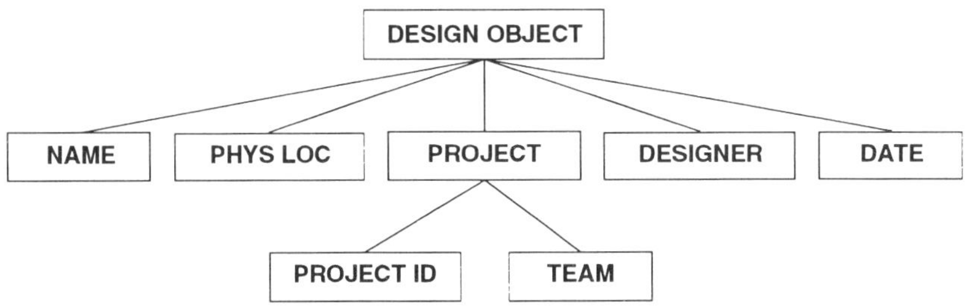

Figure 5.3 The object type Project as an attribute of type DesignObject.

Below we present some example queries to show how information about the design objects in a project projA can be retrieved:

```
GET DesignObject
   ITS Name, Designer, Date
   WHERE Project ITS ProjectID = 'projA'

GET DesignObject
   ITS Name, Date
   WHERE Project ITS ProjectID = 'projA'
   AND Designer = 'Newby'
   AND Date > 930914
```

Through the notion of project the CAD framework can provide logical distribution of design data and design activities. The framework must allow the creation and initialization of new projects in the design environment. Unrelated design activities can be separated by having them performed in different projects. Names of design objects do not clash across project boundaries and each project can be furnished in an optimal way to perform the task at hand. We have chosen to have each design object reside in one and only one project, as modeled above, since this yields a very comprehensible model for the end-user. At any time it must, however, be possible to make results of a design activity performed in one project available to a design activity that is to be performed in another project (design library facility). A mechanism that supports references of design objects across project boundaries will be presented in section 5.11.

5.4 TEAMS OF DESIGN ENGINEERS, AN EXAMPLE

According to the data schema defined above, a design object has an owner-
designer and a project has a responsible team, but the relationship among
designers and teams has not been modeled. To demonstrate how this aspect
may be included, we present the following example data schema (see Figure
5.4):

```
TYPE Team = TeamName, Mission
TYPE Project = ProjectID, Team
TYPE Membership = Designer, Team, Role
TYPE DesignObject = Name, Project, PhysLoc, Membership,
                                                       Date
```

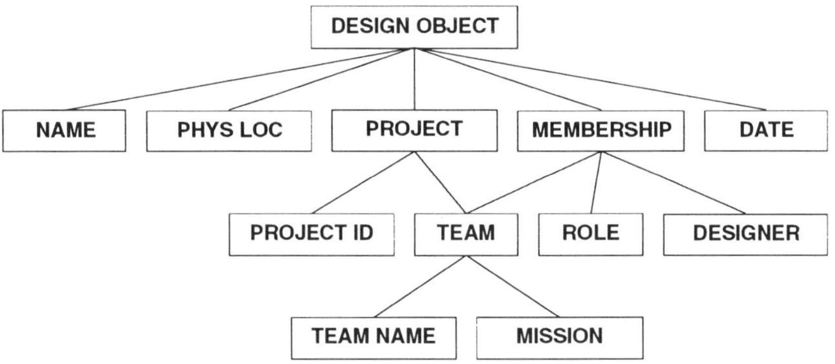

Figure 5.4   Modeling team membership of designers via the object type Membership.

A team is characterized by its *team name* and *mission*. Individual design
engineers can be made members of teams via the object type *Membership*,
with a *role* being assigned to each team member. A team may have multiple members and a design engineer may be a member of multiple teams.
Teams are non-hierarchical, that is, a team is not composed of smaller teams.
Support for hierarchical teams would require another extension of the data
schema. A design object now refers to its owner-designer via the designer's
membership of the team. A constraint that one may like to have enforced
is that in a project only members of the responsible team may own design
objects. In OTO-D this is expressed as follows:

```
ASSERT DesignObject ITS ValidMember (TRUE) =
                 Membership ITS Team = Project ITS Team
```

The team and membership information can be used to implement an *access control* strategy in the CAD framework. For example, access of a design engineer to a project may only be allowed for team members, and modifications of design objects may only be performed by owner-designers.

5.5   DESIGN TRANSACTIONS

In section 3.4 we introduced the notion of *design transaction* to control the access of design tools to design objects. A design transaction was defined as a transaction performed by a design tool on a design object (Definition 3.7), and a transaction model was presented.

We now want to include the notion of design transaction in our data schema. Reading Definition 3.7 through our OTO-D glasses, we recognize an object type *DesignTransaction* characterized by, among other things, the object types DesignObject and Tool. Other relevant information about a design transaction is the *designer* running the tool, the *date* the transaction is performed, and the *mode of access* used by the tool. The latter may for instance be used to distinguish *ReadOnly* requests from *Update* requests. This yields the following definition:

```
TYPE DesignTransaction = DesignObject, Tool, Designer,
                                   Date, AccMode
```

A design transaction is registered upon a CheckOut request issued by a design tool, after the framework kernel has decided that the tool is allowed to execute the transaction on the requested design object, as explained in section 3.4. The framework may take the access mode into account when deciding on concurrency conflicts. The registered design transaction serves as a lock on the corresponding design object.

A design transaction is terminated by a CheckIn request, with either mode commit or mode rollback. We may choose to delete the corresponding instance of the object type DesignTransaction on this occasion. A preferable solution, however, is to retain the instance while registering the *mode of completion*. For this purpose we introduce an additional attribute *ComplMode* of the object type DesignTransaction. Possible attribute values may be *Run-*

ning, *Success* and *Failed*, for running, committed and rolled-back design transactions, respectively. Switching the ComplMode value of a registered design transaction from Running to Success or Failed effectively unlocks the corresponding design object. Keeping a record of completed design transactions provides the basis for a *design transaction history* facility. Such a facility makes knowledge of which design transactions were performed on which design objects available to both the system and the end-user.

The object type *Tool* is used to register the available tools in the design environment. We define Tool as follows:

```
TYPE Tool = ToolName, Path, Possible-Opts
```

That is, a tool is characterized by its *name*, the *path* of its executable object in the file system, and the *possible options* that may be passed upon invocation. The latter may, for instance, be used by an interactive command building facility to support convenient tool invocation.

We distinguish between a *tool* and a *tool run*. A tool run is a particular run of a tool, executed at a certain date by a design engineer in a project environment, with specific options being passed on the command line. In other (OTO-D) words, we have identified a new object type *ToolRun* characterized by the attributes *Tool*, *Project*, *Used-Opts*, *Start-Date* and *Designer*:

```
TYPE ToolRun = Tool, Project, Used-Opts, Start-Date,
                                                 Designer
```

Introducing this object type in our data schema, we have to reconsider the definition of the object type DesignTransaction. Information on design transactions can now be more specific by referring to the actual tool run rather than just the tool. Thus, we refine the definition of DesignTransaction by replacing the attribute Tool by the attribute ToolRun. This also correctly reflects that during a single run of a tool multiple design transactions may be performed. The attribute Designer of DesignTransaction may now be omitted, since this attribute is available as ToolRun ITS Designer: The design engineer that executes the tool run implicitly executes all the design transactions performed during that tool run, and is administered once as an attribute of ToolRun. The dates of the individual design transactions performed in the course of a single tool run may differ. Hence, we let each design transaction have its own date. This yields the following definitions (see Figure 5.5):

```
TYPE Tool = ToolName, Path, Possible-Opts
TYPE ToolRun = Tool, Project, Used-Opts, Start-Date,
                                          Designer
TYPE DesignTransaction = DesignObject, ToolRun, Date,
                                     AccMode, ComplMode
```

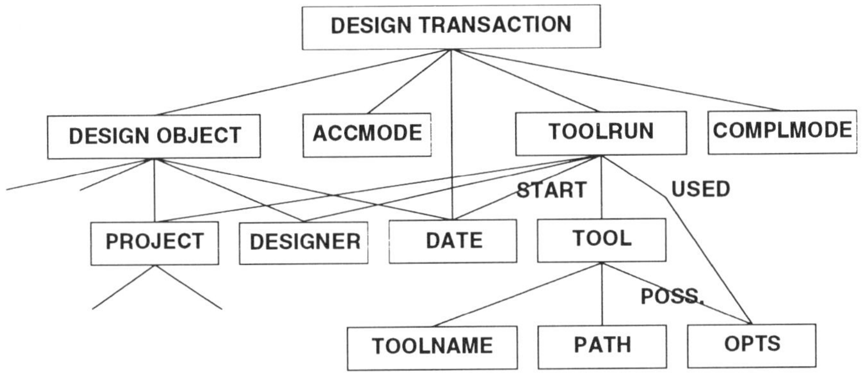

Figure 5.5 The object types Tool, ToolRun and DesignTransaction included in the data schema.

We see that the recognition of the relevant object types has yielded an aggregation hierarchy Tool-ToolRun-DesignTransaction. This aggregation hierarchy offers different *levels* at which relevant information may be registered, as appeared from the discussion about the Designer and Date attributes.

Note that we do *not* ASSERT the constraint that for a particular tool run initiated in a particular project environment only design transactions may be performed on design objects from that project environment. In other words, tool runs may cross project boundaries.

The data schema of Figure 5.5 models a single global set of tools, and does not allow for different projects to have different sets of selected tools. Explicit selection of tools for use in projects could be modeled by introducing an object type ToolSelection that relates tools to projects. ToolRun would then refer to ToolSelection instead of Tool, implying that a tool has to be selected first for use in a project before it can be run. This would yield the following type definitions:

```
TYPE Tool = ToolName, Path, Possible-Opts
TYPE ToolSelection = Tool, Project
TYPE ToolRun = ToolSelection, Used-Opts, Start-Date,
                                              Designer
TYPE DesignTransaction = DesignObject, ToolRun, Date,
                                   AccMode, ComplMode
```

Note that this data schema modification does not affect the definitions of
the object types Tool and DesignTransaction. We will address the topic of
project-specific tool selections again in section 5.12, where we will see an
elegant solution emerge in the context of design flow management.

Based on the data schema of Figure 5.5, a wealth of information can be
made available to the design engineer. This includes information about the
available tools, the *runs of tools* and the *accesses on design objects* that are
being performed or have been performed during runs of design tools. The
following example queries demonstrate the retrieval of valuable information
on the basis of this data schema:

```
GET ToolRun
  ITS Used-Opts, Designer, Start-Date
  WHERE Tool ITS ToolName = 'Dracula'
  AND Start-Date >= 931027
```

The above query tells which runs were performed with the Dracula tool
since October 27, 1993. For each run it presents the options used, the design
engineer who performed the run, and the date the run was started.

```
GET DesignTransaction
  ITS DesignObject ITS Name,
    DesignObject ITS Project ITS ProjID,
    ToolRun ITS Designer
  WHERE ToolRun ITS Tool ITS ToolName = 'Dali'
  AND ComplMode = 'Running'
```

This query finds out which design objects are currently being accessed by
the tool Dali and which design engineer is running it. In a similar way
information can be retrieved about running transactions on a particular design
object. The framework kernel itself may use this kind of information to
decide on concurrency matters upon a CheckOut request by a design tool.
A design engineer who needs access to a design object can determine who

currently holds it, enabling him to negotiate for its early return.

```
GET DesignTransaction
   ITS ToolRun ITS Tool ITS ToolName, Date,
                                        ToolRun ITS Designer
   WHERE DesignObject ITS Name = 'FlipFlop'
   AND DesignObject ITS Project ITS ProjID = 'projC'
```

The above query retrieves which tools were run when and by whom on the design object FlipFlop in project projC. This design transaction history can inform the design engineer about the verification status of the particular object.

5.6 VIEW TYPES

An inherent aspect of electronic design, as well as other design-oriented application areas, is the use of multiple *levels of abstraction* at which a design can be represented. The CAD framework has to support the management of the multiple representations of a design that may be entered at these different levels. The different representations of a design are often referred to as the *views* of that particular design. For example, in electronic design one can have for a particular design a layout view at the mask-layout level, a circuit view at the transistor level, and a RTL view at the register-transfer level. We refer to the abstraction level of a design representation as the *view type*.

Design representations can be entered by the design engineer, but they can also be derived from each other (semi-)automatically by synthesis tools or analysis tools. Typically, automatic synthesis tools generate descriptions at a lower level of abstraction from descriptions at a higher level. The reverse is typically performed by analysis tools for purposes of design verification; a higher level description is derived to verify whether the corresponding lower level design description satisfies certain rules or demonstrates required behavior. One could argue that derived design representations do not have to be stored explicitly since they can be (re-)generated dynamically on an "on demand" basis, that is, when needed in subsequent design activities. A more practical approach, however, is to allow derived design representations to be stored explicitly when so desired, and have them managed with the help of the CAD framework. The derivation process may be computationally intensive and the design engineer may want to retain derived design representations for future use. The CAD framework has to provide support for organizing the multiple design representations as well as for maintaining their relationships.

We extend our data schema to support the organization of multiple design representations. According to Definition 3.6, a design object is managed as a single logical unit. Multiple design representations, which are allowed to exist individually, are hence to be organized as multiple design objects. A single design object contains a description of a design at a particular level of abstraction. We extend the definition of the object type DesignObject with the classifying attribute *ViewType*, to label each design object with the view type of the design representation it contains (see Figure 5.6).

```
TYPE DesignObject = Name, ViewType, Project, PhysLoc,
                                              Designer, Date
```

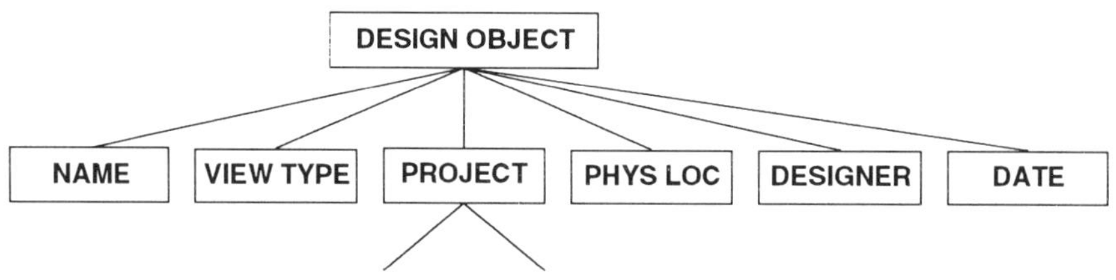

Figure 5.6   The object type DesignObject with the classifying attribute View-Type.

In section 5.3 the scope of the name of a design object was said to be the project it resides in. The attribute ViewType allows us to distinguish between design objects having the same name but different view types. Hence, we reduce the scope of the name of a design object. A system-wide identification of a design object is now given by the triplet <project, view type, name>. Within a project environment the view types provide *name spaces*, i.e. categories in which design objects can be uniquely identified by their name. This allows different design representations to co-exist under the same name.

It is not up to the CAD framework to determine which view types are supported in a design environment. New design methodologies and their associated representations are still evolving. According to the above data schema, design objects are simply *labeled* with their view type, but the actual set of view types has not been fixed. For an actual design environment we must allow the set of supported view types to be configured.

An interesting modeling question is whether the framework should allow a design object to be of multiple view types. This would permit more extensive labeling of design objects that can serve multiple purposes in the design process. However, if a design object is allowed to be of multiple view types, then the name spaces provided by the view types are no longer disjunct: Individual objects may appear in multiple categories. This blurs the clear design object classification and identification scheme defined above.

Is there a great need for multi-view design objects? Most design tools have been designed to synthesize or verify design descriptions at a particular level of abstraction. A well-integrated design environment must offer a well-defined set of abstraction levels at which designs can be represented, with each level being supported by a set of design tools for synthesis and verification at that level. This is in line with the framework approach of having a *modular* set of design tools integrated on top of a CAD framework, communicating via standardized design representation formats. The need for multi-view design objects may arise through the introduction of a tool that generates design descriptions covering multiple levels from the existing set.

Having only single-view design objects, one must decide upon introduction of a new type of design description whether to combine it with an existing view type, or to introduce a new view type, or to have the new descriptions stored as multiple related design objects of different (possibly existing) view types. We note that other classification mechanisms may further help in organizing design information. Additional labeling of design objects can be supported by defining one or more extra attributes for the object type DesignObject. For a particular view type these attribute(s) may be used to further classify the design objects of that view type. Such a *sub-typing* facility can be helpful in integrating particular application environments and does not interfere with the design object identification scheme. Another thing to remember is that proper design flow management facilities maintain the history and state of design objects, and control which tool operations data may be involved in next, as configured in the design flow. Design flow management inherently provides data classification based upon the operation of tools on data. Given these additional classification capabilities, we feel that design engineers are served best by a simple but effective view type mechanism for purposes of data organization and identification. We therefore classify design objects by means of a single view type attribute.

According to the data schema of Figure 5.6, each design object carries its own name and view type. This implies, for instance, that design objects can be renamed individually. A slightly different approach would be to introduce a new object type to represent the set of design objects having the same name but different view types. We term this object type *Cell*, which is consistent with the terminology employed by e.g. CFI [CFI93, CFI92b], JCF [JCF92, LJ92], EDIF [EDI90] and Oct/VEM [HMSN86]. We define cell as follows:

Definition 5.2 *A* cell *is a named piece of design for which one or more representations exist.*

As we said above, these representations are also called views. We refine our data schema as follows (see Figure 5.7).

```
TYPE Project = ProjectID, Team
TYPE Cell = Name, Project, Designer
TYPE DesignObject = Cell, ViewType, PhysLoc, Date
```

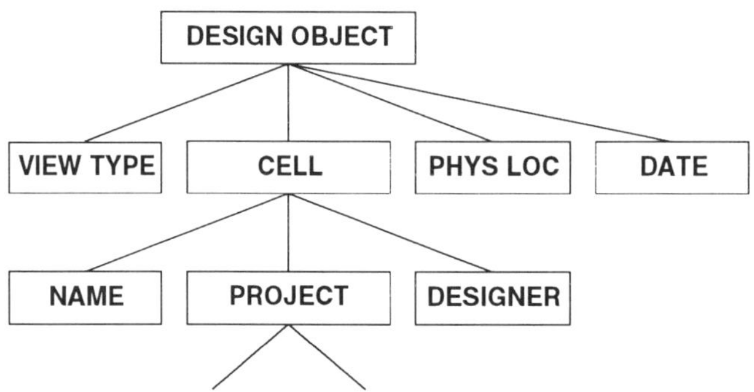

Figure 5.7 Design objects are organized as views of named cells. The object type Cell has the attribute Name, and the object type DesignObject has the attributes Cell and ViewType.

According to this data schema, design objects are organized in named cell containers owned by designers. Design objects are further characterized by their view type, physical location and date of construction. A synonym of DesignObject would be View, as the design objects are the views of cells. Relatability guarantees that each design object belongs to a cell. Multiple design objects may belong to the same cell. As an additional constraint we require at least one design object, or view, to belong to a cell:

```
ASSERT Cell ITS NumberOfViews (1..*) =
                        COUNT DesignObject PER Cell
```

Note that Designer is an attribute of Cell, that is, ownership is handled at the cell level. If ownership is to be arranged per design object, then Designer should be made an attribute of DesignObject.

The refinement of the data schema from the one in Figure 5.6 to the one in Figure 5.7 does not affect the identification scheme of design objects. A system-wide identification of a design object is still given by the triplet <project, view type, name>. The principal difference is that naming is now performed at the cell level, rather than at the level of the individual design objects. Further, this data schema refinement affects neither the definition of the object type Project nor the definitions of object types that (will) have DesignObject as an attribute.

We emphasize that a cell should merely be seen as a container object that clusters design objects of different view types under a common name. It is *not* intended to imply equivalence among its member design objects. Above we briefly discussed the derivation of one design representation from the other. We now conclude that both the original and the derived design representation are to be handled as individual design objects, each of a particular view type. The CAD framework has to provide support for maintaining the relationship between these design objects. The support for explicit equivalence relationships between design representations will be addressed in section 5.9.

5.7   VERSIONING

Design is iterative and tentative. Design engineers often take their design descriptions through several refinements and explore different design alternatives seeking for the best solution. To support evolutionary and exploratory design, the CAD framework must allow the design engineer to have multiple versions of a piece of design and help him in managing these versions. It must maintain a derivation history and support selection and use of individual versions.

The CAD framework has to maintain additional information for organizing the versions of a design. We therefore have to extend the data schema to represent new types of information. Let's try to recognize the relevant object types. We saw that the framework must allow multiple versions of a design. That is, we have the *individual versions* and the *sets-of-versions* to which the versions belong.

A design object is our smallest logical unit. Hence, a design object is an individual version, rather than something we have versions of. Also remember that tools perform design transactions on design objects; the state of design evolves through operations on individual design objects. Therefore, we make the design objects our version entities. To yield a comprehensible model for the end-user, we choose a design object to be a version of one and only one set-of-versions. That is, version sets do not overlap. Per set-of-versions we distinguish the individual version design objects by means of a *version number*. In addition each version has a *version status*. This status can be used to classify versions, for example, for purposes of selection.

The next question is how to organize version design objects into sets-of-versions. That is, which object type is to represent the sets-of-versions. We propose two modeling alternatives. The OTO-D diagrams are presented in Figure 5.8.

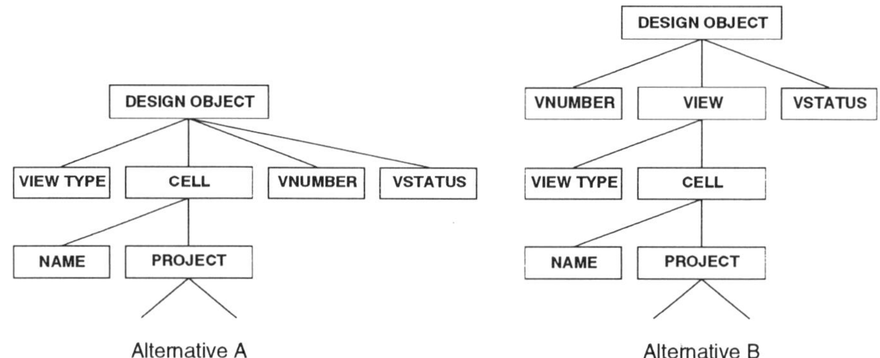

Figure 5.8   Two modeling alternatives for organizing version design objects.

According to Alternative A in Figure 5.8 each design object carries its own view type, and design objects of different view types may relate to the same cell. This makes cell a multi-view version-set, and suggests that versioning of a design at a particular level of abstraction interferes with versioning of the design at other levels of abstraction.

Alternative B in Figure 5.8 introduces the object type View as a single-view version-set. A view relates to a cell and carries a view type. A design object is a numbered version of a view. Multiple design objects may belong to a view as its different versions, sharing the view type as DesignObject ITS View ITS ViewType. Alternative B orthogonalizes the aspects of versioning and multi-view design in an attractive way.

We select Alternative B, and correspondingly define view as follows:

Definition 5.3 *A* view *is a named piece of design at a particular level of abstraction for which one or more versions exist.*

Thus, a view is a representation of a cell of which we can have different versions. Use of the term view for a representation of a cell is consistent with the terminology employed by e.g. CFI [CFI92b], JCF [JCF92], EDIF [EDI90] and Oct/VEM [HMSN86]. CFI also uses the term *cell view* [CFI93].

Reflecting our choices in OTO-D type definitions, we come to the following data schema.

```
TYPE Project = ProjectID, Team
TYPE Cell = Name, Project, Designer
TYPE View = Cell, ViewType
TYPE DesignObject = View, Vnumber, Vstatus, PhysLoc,
                                                    Date
```

This data schema constitutes the aggregation hierarchy Project-Cell-View-DesignObject. Relatability guarantees that each design object belongs to a view, and that each view belongs to a cell. Multiple design objects may belong to the same view, and multiple views may belong to the same cell. The cell carries the name and the view carries the view type. A view is uniquely identified by the triplet <project, view type, name>. Versioning introduces an extra level in referencing the individual design objects. A design object is uniquely identified by ITS View *plus* an additional version number: <project, view type, name, version number>. As additional constraints we require at least one design object to belong to a view and at least one view to belong to a cell:

```
ASSERT View ITS NumberOfVersions (1..*) =
                            COUNT DesignObject PER View
ASSERT Cell ITS NumberOfViews (1..*) =
                            COUNT View PER Cell
```

The aggregation hierarchy offers different levels at which attributes may be attached. For example, each design object has its own date of construction and physical location. We have made Designer an attribute of Cell, implying that ownership is handled at the cell level. The design objects of the views of a cell share the owner-designer as DesignObject ITS View ITS Cell ITS Designer. Alternative solutions would be to arrange ownership per view or per design object, which are modeled by making Designer an attribute of View or DesignObject, respectively.

The following example queries demonstrate the possibilities for information retrieval on the basis of the derived data schema:

```
GET View
   ITS ViewType
   WHERE Cell ITS Name = 'FlipFlop'
   AND Cell ITS Project ITS ProjID = 'projB'
```

The above query tells which views we have for a cell named FlipFlop in project projB.

```
GET View
   ITS Cell ITS Name
   WHERE ViewType = 'VHDL'
   AND Cell ITS Project ITS ProjID = 'projB'
```

This query retrieves the names of the cells in project projB for which we have a VHDL description.

```
GET DesignObject
   ITS Vnumber, Vstatus, Date
   WHERE View ITS ViewType = 'circuit'
   AND View ITS Cell ITS Name = 'latch'
   AND View ITS Cell ITS Project ITS ProjID = 'projB'
```

This query retrieves information about the versions of the circuit description of the cell named latch in project projB. For each version it retrieves the version number, the version status and the date of construction.

According to the data schema, we have a collection of numbered version design objects per view. Yet, a simple linear numbering scheme is not sufficient to capture the derivation history of these versions. This may be a non-linear evolution, since multiple versions may be derived from the same original version. Also, we can think of cases where a new version is derived from multiple original versions. The version derivation history of the versions of a view can be represented by a *version derivation graph*, which may contain branches as well as merges.

We could try to adopt a more complex version numbering scheme in order to capture version derivation histories. However, this may get quite complicated, while it also interferes with the design object identification scheme. We therefore choose to model the *version derivation relationships* between design objects explicitly via a new object type *VersionDerivationRel*. A version derivation relationship is characterized by the *original design object* and the *derived design object* derived from it. Other information could be the *tool run* by which the derivation was performed. This information, however, is already available via the design transaction that is administered upon the derivation. We do not include it again, since this would introduce redundancies. This leads to the following data schema (see Figure 5.9):

```
TYPE Project = ProjectID, Team
TYPE Cell = Name, Project, Designer
TYPE View = Cell, ViewType
TYPE DesignObject = View, Vnumber, Vstatus, PhysLoc,
                                                Date
TYPE VersionDerivationRel = Original-DesignObject,
                                Derived-DesignObject
```

Note that role attributes have been used in the definition of the object type VersionDerivationRel, to distinguish the DesignObject attributes. The object type VersionDerivationRel is to represent version derivation relationships among the versions of a view. We, therefore, impose the constraint that the original and derived design object belong to the same view:

```
ASSERT VersionDerivationRel ITS ValidRel (TRUE) =
        Original-DesignObject ITS View =
                        Derived-DesignObject ITS View
```

Other types of derivation relationships, possibly between design objects of different view types, will be covered in section 5.9.

The example version derivation history of a view depicted in Figure 5.10 can be supported by the CAD framework on the basis of the data schema of Figure 5.9. Both branching and merging are supported.

Our version handling concept has similarities with the work of Katz as pre-

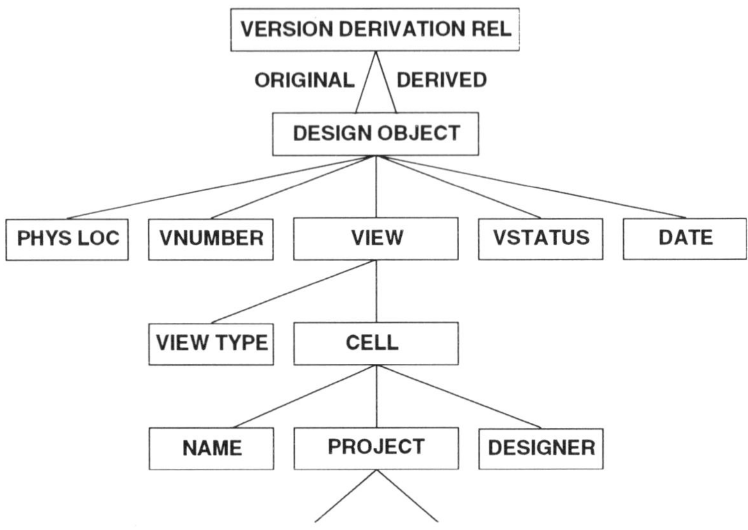

Figure 5.9   Data schema extensions to support versioning.

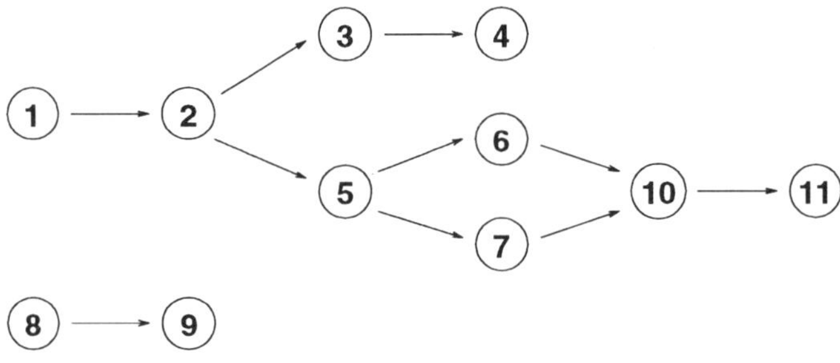

Figure 5.10   Example version derivation history supported by the defined data schema. Numbered version design objects are related by binary version derivation relationships.

sented in [KAC86] and [KBC$^+$87]. Katz also organizes numbered design objects (called representation objects) in graph-like version derivation histories per view type. The Jessi-Common-Framework, on the other hand, performs versioning at the level of multi-view cell objects.

The data schema of Figure 5.9 defines how version design objects are organized in views and cells and defines that version derivation relationships between design objects are used to capture version derivation histories. It specifies the global aspects of version management, but leaves the detailed functionality largely open. For example, we have not yet defined which version statuses are used by the system and how versions are treated depending on their status. The version status may indicate whether a version is 'released' or 'in-progress', and may be used to control version mutability and to direct version selection. See e.g. [CK88], [Kat90] and [OTC93] for more detailed discussions on version management.

5.8 HIERARCHY

The most obvious way to master the inherent complexity in the design of large integrated systems is by structuring the design in a hierarchical way. The system under design is decomposed into several smaller sub-systems, which in turn can be decomposed into even smaller sub-sub-systems, etc. The structure of a hierarchical design can be typified as a *directed acyclic graph*, with the vertices representing the (sub-)systems and the edges representing the hierarchical relationships between the (sub-)systems. Note that the term hierarchy is often connected to tree-like structures. We use it to denote graph-like structures, which allow us to also represent the reuse of sub-systems. The example in Figure 5.11 represents the hierarchical decomposition of a design A.

A *hierarchical relationship* is a directed *reference* from a compound or *parent* design description to a component or *child* design description. It represents the use of the child design description as a component in the parent design description. We say that the child design description is *instantiated* in the parent design description, and the actual use of the child design description in the parent is called an *instance* of the child design description. A design description can be instantiated by multiple different parent design descriptions. For example, see Figure 5.11 where E is instantiated by both A and B. Also, a design description can be instantiated multiple times by one and

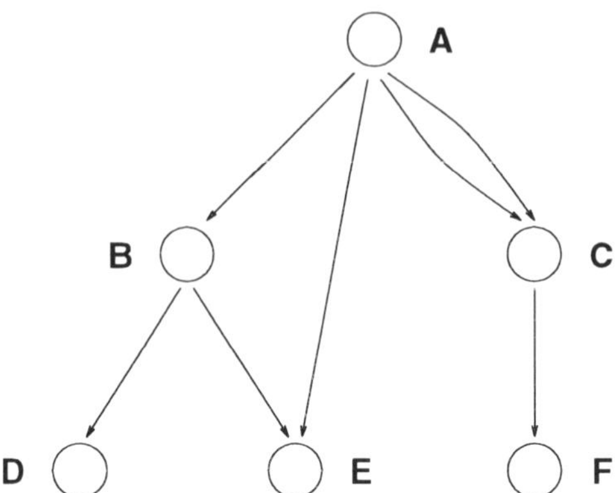

Figure 5.11 Hierarchical decomposition of a design A.

the same parent design description. For example, in Figure 5.11 sub-system C is instantiated two times in A. For each instantiation there is information unique to the particular instance of the child in the parent. This instance-specific information describes how the child design description is actually used in the parent. We refer to it as the *constructor*.

In the design database a hierarchical design is to be organized according to its graph-like structure. That is, each design description is to be stored once as an individual entity, and the relationships between design descriptions are to be stored together with their respective constructors. Design tools must be allowed to access individual design descriptions and to traverse both up and down the design hierarchy. They can use the constructors to compute the actual instances, thereby instantiating the directed acyclic graph into a tree of instances. The content of the constructor depends on the type of design description.

We now return to our data schema, to extend it for representing hierarchical design structures. A vertex in a hierarchy graph represents a design description at a particular level of abstraction. This corresponds to our design object, modeled by the object type DesignObject. A hierarchical relationship is characterized by its *parent design object*, its *child design object*, and the *constructor* for the particular instantiation. There is a one-to-one correspondence between hierarchical relationship and instance. We allow an *instance name* to be attached to each hierarchical relationship, to be used for identification of the different instances within a parent design description. This brings us to the following extension of the data schema (see Figure 5.12):

```
TYPE HierarchyRel = Parent-DesignObject,
                 Child-DesignObject, InstName, Constructor
```

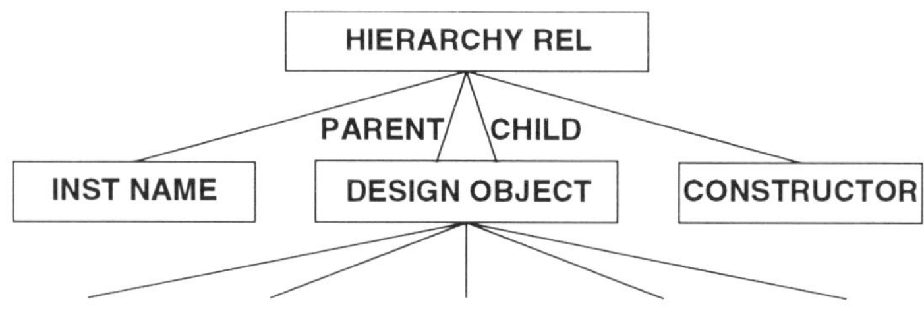

Figure 5.12 Data schema extension to support hierarchical design.

According to this OTO-D type definition, the information concerning an instance of a child design object in a parent design object is aggregated into a separate entity, as is also proposed in e.g. [CFI92b] and [SW92]. The two DesignObject attributes each have their own role, denoted by the prefixes *Parent* and *Child*: role attributes. The schema correctly reflects that an individual design object may be involved in zero or more hierarchical relationships, both as a parent and as a child design object. Relatability guarantees that a hierarchical relationship can exist only with both a parent and a child design object.

The data schema of Figure 5.12 permits traversal both up and down the design hierarchy. The following queries illustrate how the parent design objects and the child design objects of version # 3 of the circuit view of the cell register can be retrieved. In other words, which design objects use the register as a component, and which design objects does the register itself use as a component. Notice the similarity of the two requests.

```
GET HierarchyRel
  ITS InstName,
    Parent-DesignObject ITS View ITS Cell ITS Name,
    Parent-DesignObject ITS Vnumber
  WHERE Child-DesignObject ITS View ITS Cell
                                ITS Name = 'register'
  AND Child-DesignObject ITS View ITS ViewType =
                                'circuit'
  AND Child-DesignObject ITS Vnumber = 3
```

```
GET HierarchyRel
  ITS InstName,
    Child-DesignObject ITS View ITS Cell ITS Name,
    Child-DesignObject ITS Vnumber
  WHERE Parent-DesignObject ITS View ITS Cell
                                    ITS Name = 'register'
  AND Parent-DesignObject ITS View ITS ViewType =
                                    'circuit'
  AND Parent-DesignObject ITS Vnumber = 3
```

5.9 EQUIVALENCE RELATIONSHIPS

In section 5.6 we discussed the derivation of design descriptions from other design descriptions, and the requirement put upon the CAD framework to provide support for maintaining the relationships between these design descriptions. We call such relationships *equivalence relationships*, since the original design description and the derived design description are equivalent in certain respects. Equivalence may also occur without derivation of one design description from the other. For example, when two design descriptions, of possibly the same view type, are *proven* to be equivalent. We will use the term equivalence in quite a broad sense, making few assumptions about the *kind* of relationship between design descriptions. This will yield a generic mechanism that can be used for many different purposes in the construction of application environments. It will allow design tools to register the relationships that they consider to be valid, for use in subsequent tool runs or to inform the design engineer.

We extend our data schema to support equivalence relationships. One way could be to introduce *equivalence-sets*, being sets of design objects shown to be equivalent. An individual design object can then be made a member of a set when so desired. This solution raises the question of *transitivity* of equivalences: if A is equivalent to B and B is equivalent to C, is A equivalent to C? Equivalence relationships are the reflection of complex application-specific derivation and proof procedures of which the framework by itself has no specific knowledge. Critical information about equivalence transitivity could be fed to the framework through configuration, to allow the framework to automatically create, join and split equivalence-sets. The main drawback, however, is the lack of simplicity, both for application builders and for design engineers. In [KAC86] Katz employs the concept of equivalence-sets.

An alternative solution is to model equivalence relationships as binary relationships between design objects. This yields a simple and flexible approach towards the management of equivalence information. It allows applications to relate design objects on a pairwise basis. The framework simply administers equivalence relationships as they occur, without any premature assumption about their nature, and leaves their interpretation to the applications. Some examples: For a run of a synthesis tool, the synthesized design description is simply related to the original design description that served as input. For a netlist comparison tool, the two netlists are related if they have proven to be structurally equivalent. Upon circuit simulation, the simulation output is related, by means of two equivalence relationships, to both the circuit that was simulated and the stimuli that were used to drive the simulation. We also note that for the representation of the equivalence information to the end-user the most appropriate (graphical) form can still be chosen freely.

We define the binary equivalence relationship as follows (see Figure 5.13):

```
TYPE EquivalenceRel = Source-DesignObject,
                      Target-DesignObject, ToolRun, Class
```

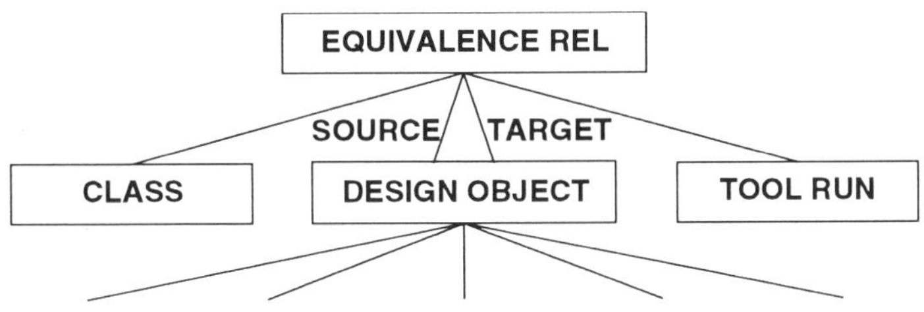

Figure 5.13   Data schema extension to support equivalence relationships.

The ToolRun attribute is used to register the tool run during which the relationship was established. The Class attribute can be used to typify relationships. Two design objects of particular view types may be equivalent in different respects. The Class attribute allows the different kinds of equivalences to be distinguished. An individual design object may be involved in zero or more equivalence relationships; the schema supports many-to-many relationships among design objects. Relatability guarantees that an equivalence relationship can exist only if both the source and the target design object exist.

5.10   THE INTERPLAY AMONG VERSIONING, HIERARCHY AND EQUIVALENCE

5.10.1   Introduction

In the previous sections we addressed three key data management topics: 1) versioning, 2) hierarchy and 3) multiple representations and equivalence. These are sometimes referred to as the *three dimensions of data management* [KAC86]. So far, data schemas have been defined separately for these topics, and it is now time to perform an additional validation step. We therefore wonder what the interplay between the defined concepts is:

- How do they affect each other?

  For example, if a Register has a hierarchical relationship with a FlipFlop, which version of the FlipFlop actually gets instantiated?

- Does the interplay yield built-in structural constraints?

  For example, are equivalent objects constrained to have isomorphic hierarchical decompositions?

- How is consistency handled?

  What about propagation of changes and enforcement of consistency constraints?

Below we present the data schema in which the principal modeling aspects of the three topics have been combined (see Figure 5.14):

```
TYPE Project = ProjectID, Team
TYPE Cell = Name, Project, Designer
TYPE View = Cell, ViewType
TYPE DesignObject = View, Vnumber, Vstatus, PhysLoc,
                                                  Date
TYPE VersionDerivationRel = Original-DesignObject,
                                    Derived-DesignObject
TYPE HierarchyRel = Parent-DesignObject,
            Child-DesignObject, InstName, Constructor
TYPE EquivalenceRel = Source-DesignObject,
                Target-DesignObject, ToolRun, Class
```

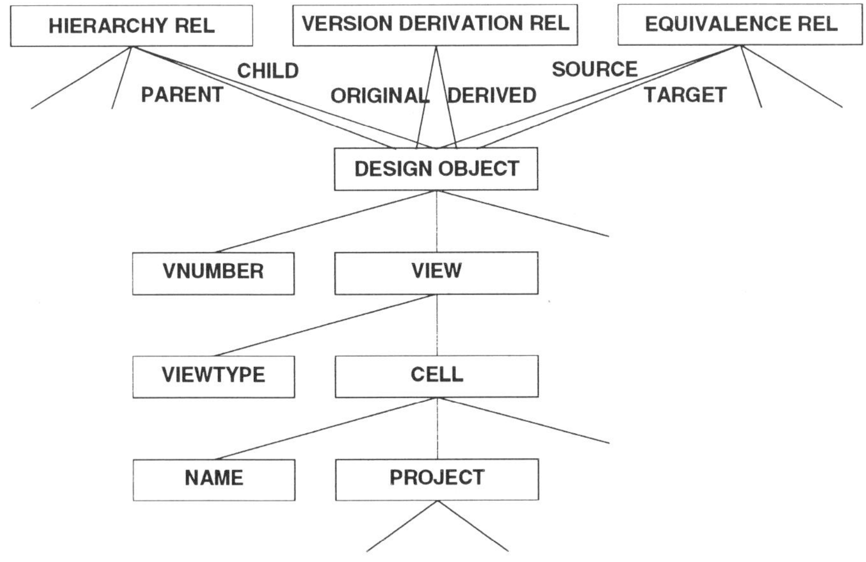

Figure 5.14 The combined data schema for the key data management topics of versioning, hierarchy and multiple representations and equivalence.

A general observation from Figure 5.14 is that both hierarchical and equivalence relationships relate *individual versions* (as do the version derivation relationships, of course). That is, for each version the system explicitly administers to which other versions it relates. In the following sub-sections we will discuss the interplay between the defined concepts per two topics.

5.10.2 Equivalence and Versioning

Equivalence relationships relate individual versions, which is exactly what we want. Design objects stand for the design descriptions that may be derived from each other, so these are the entities to relate. No implicit equivalence is suggested amongst the different views of a design, as is e.g. done in [Kat83]. The version mechanism simply collects design objects in single-view version-sets, the views, and all equivalences are administered explicitly by the equivalence mechanism. There is no equivalence implied by the organization of design objects as versions of views of specific view types. Also there are no constraints on the equivalences that a version is allowed to have, based on the equivalences of its 'brother' versions of the same view.

A distinction between version derivation relationships and equivalence relationships is that the former are generated by the framework itself, in order to maintain version derivation histories, and the latter are generated by design tools in order to register application-specific relationships among design objects. The example in Figure 5.15 illustrates the interplay among versioning and equivalence.

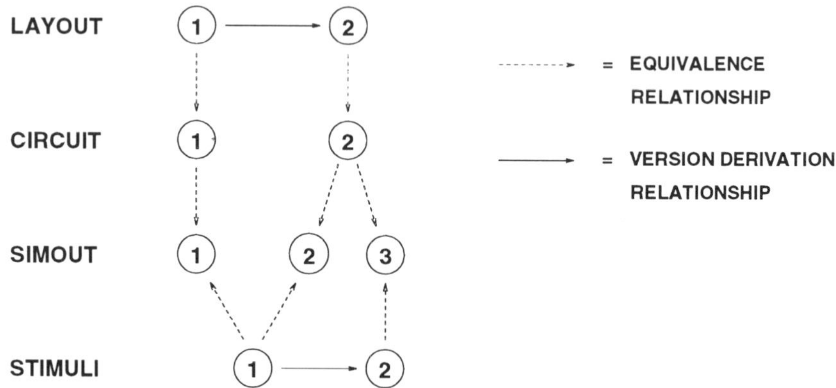

Figure 5.15   Example of the interplay among versioning and equivalence.

In this example, equivalence relationships have been registered upon derivation of circuit objects from layout objects and upon simulation of circuit objects with stimuli objects yielding simulation output objects. For example, from the equivalence relationships in Figure 5.15 we learn that circuit # 2 has been derived from layout # 2 and has been simulated twice with different stimuli. The framework has administered version derivation relationships upon the creation of new versions of the layout and the stimuli. The version derivation relationships and equivalence relationships do not constrain each other. We conclude that the schema of Figure 5.14 offers the required flexibility for equivalence relationships with respect to versioning.

5.10.3   Hierarchy and Versioning

The interplay of hierarchy and versioning brings us to the *binding* problem: If a component, say FlipFlop, is referenced hierarchically, to which version of the FlipFlop does the reference get bound? There are two types of binding: *static* and *dynamic* [KAC86, OTC93]. Static binding implies that a hierarchical relationship is explicitly directed to a specific version. Dynamic binding implies that the selection of a specific version is deferred until the time of usage. In our view, a CAD framework must support at least static binding. A design engineer must be allowed to fix a hierarchical relationship on a specific version of a view. For example, a Register object may work perfectly well with a particular FlipFlop. When newer FlipFlops are created for use in new Registers, the design engineer may still wish to retain the older Register - FlipFlop pair as a consistent configuration. In addition, a CAD framework may offer dynamic binding as an optional facility, for example, to be used in initial design stages.

The data schema of Figure 5.14 is geared towards static binding, since versions get related explicitly, but dynamic aspects are not excluded. For example, on the basis of this data schema the CAD framework can still offer a facility to re-direct hierarchical references when a new version of a component becomes available. More explicit support for dynamic binding could be offered by extending the data schema with the following object type:

```
TYPE DynamicHierRel = Parent-DesignObject, Child-View,
                                   InstName, Constructor
```

This allows dynamic hierarchical references that relate to the generic component view rather than one of its specific versions. Upon tool accesses, such a dynamic reference must be dereferenced to a specific version of the identified view. It is important to realize that dynamic binding makes design activities more error prone, as the *creation* of a new version by itself may imply *usage* of this version in the system under design. This (implicit) change also invalidates previously derived verification results, as the inclusion of the new version may affect the implied behavior of the system under design in a critical way. Making this happen beyond control of the design engineer is often not desirable.

We are in favor of a strategy where the inclusion of a new version into a design is separated from its creation: A new version of a component must be allowed to be created and verified before it is actually going to be used. We define a framework function that allows one to explicitly include a new version into a design hierarchy. We term this function the *install operation*. The install operation is to be used by design engineers and smart design tools to explicitly propagate a change in the design hierarchy when they feel that the new version is to replace an older version. Ease of use and possibilities for semi-automatic activation of the install operation are critical to offering convenient design procedures to the end-user. The install operation will be further detailed in sub-section 5.10.5.

The proposed interplay of versioning and hierarchy offers increased flexibility, in particular when operating with multiple users on hierarchical designs. It permits flexible concurrency control strategies, as previous versions can be browsed without regard to in-progress update transactions on experimental versions that still have to be verified.

5.10.4 Hierarchy and Equivalence

We wonder whether hierarchical relationships do somehow constrain the equivalence relationships that can be established between design objects, and vice versa. For example, it is generally acknowledged that the constraint of identical hierarchical decompositions across equivalent design descriptions yields an unacceptable inflexibility. In our approach the concepts of hierarchy and equivalence have not been intertwined. In the data schema (Figure 5.14) there are no dependencies between the object types HierarchyRel and EquivalenceRel. They both relate design objects on a pairwise basis, without constraining each other. Figure 5.16 gives an example of the hierarchical multi-view structure of a design based upon the presented data schema.

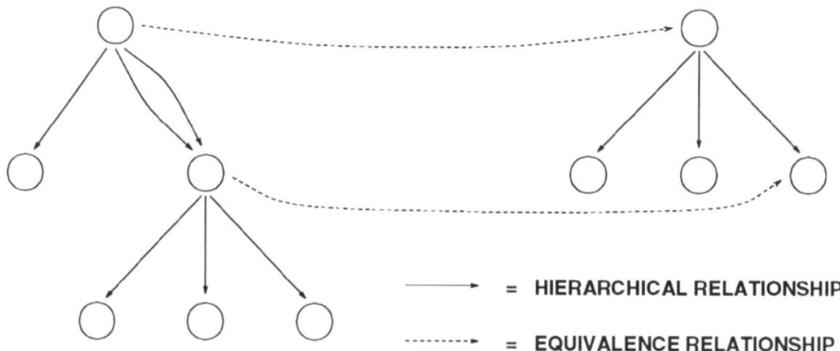

Figure 5.16 Example of the hierarchical multi-view structure of a design.

Design objects are related either 'vertically' by hierarchical relationships or 'horizontally' by equivalence relationships, without any mutual constraints. This is sometimes called the *hierarchy multi-view matrix* [Dew86]. Structure, where it exists, can be exploited without constraining design tools or design engineers.

5.10.5 Change Propagation and Consistency Constraints

The issue of change propagation and consistency maintenance in the hierarchical multi-view context has not been covered extensively in the literature. Some publications addressing this topic are [Kat85a], [BK85], [CK88], [Kat90], and [OTC93].

For consistency purposes we take the approach that an update of a design description is never done in-place, but always through the creation of a new design object. That is, upon an Update design transaction, the transaction mechanism starts from an existing design object and produces a new design object. This new design object becomes a new version of a view. On this occasion the framework may optionally remove the existing version from which the new version was derived, thereby mimicking an *overwrite*, but only if this older version is not being used somehow (e.g. referenced hierarchically). The new version is a 'clean' version which carries no invalid verification results. These will have to be re-derived, possibly leading to equivalences relating the new design object to other (new) design objects.

As already mentioned, a new version can be included into a design hierarchy, or configuration, through the *install operation*, which will 'substitute' the new version for the one currently being used in this design hierarchy. This change has to be propagated upward in the hierarchy, as the inclusion of the new version may critically affect the behavior of the referencing systems under design. As a consequence, equivalences of design objects that (indirectly) reference the new version have to be invalidated.

In our approach, when a new child version is to be installed in a parent, a new parent version is created that references the new child version rather than the existing child version. For the other components, the new parent version simply references the existing child versions. This is consistent with the procedures for change propagation described in [Kat85a], [Kat90] and [OTC93]. The new parent version optionally overwrites the existing parent version. This procedure may be performed recursively, upward in the design hierarchy. The scope of the propagation must be under control of the design engineer. He must be able to specify along which branches propagation must be performed and how far up in the design hierarchy it must proceed. The install operation is illustrated by the example in Figure 5.17.

Note that the new design objects that (indirectly) reference the new version have no invalid equivalences related to them. An extension would be to allow

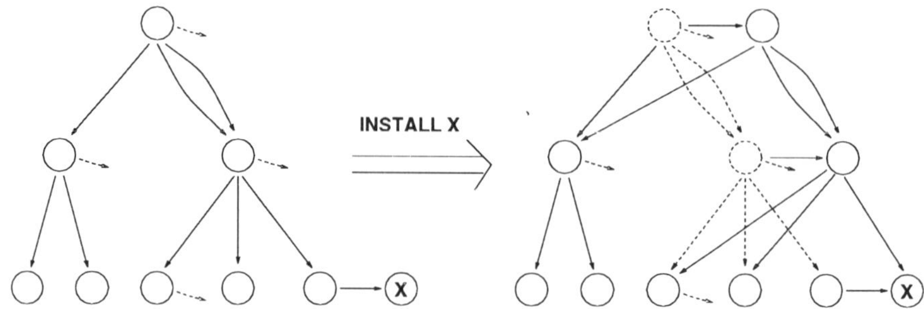

Figure 5.17 Propagation of changes through the install operation. Version 'X' is installed recursively, leading to a new configuration that includes version 'X'.

the new parent version(s) to inherit verification results that are guaranteed to be still valid. This must be configurable for the different types of verification results in an application environment.

With static binding, identification of a design object implicitly identifies its complete design hierarchy, or configuration. The presented strategy for change propagation does not modify design objects in-place. Updates yield new design objects, which optionally overwrite existing design objects, if possible. This implies that a design hierarchy is implicitly 'frozen' at the moment a tool initiates a design transaction on its root object. As a consequence design tools do not have to adhere to complex locking procedures in order to operate correctly in a *multi-user* environment. The install operation employs the locking facilities of the design transaction mechanism to compete with tools for access to design objects.

5.10.6   Discussion

One principal conclusion from the data schema that we have derived for the key data management topics (Figure 5.14) is that we have separated relationship matters from the organization and identification of design objects. The aggregation hierarchy Project-Cell-View-DesignObject defines how design objects are organized hierarchically in view, cell and project containers, and defines the identification scheme for design objects. However, this organization & identification of design objects does not imply any particular relationships among design objects. For example, no equivalence is implied by the way design objects are organized in views and cells. Such relationships are handled explicitly. The object types VersionDerivationRel,

HierarchyRel, and EquivalenceRel allow binary relationships to be established between uniquely identified design objects. These relationships are independent of the way the design objects are organized in views and cells (additional constraints may of course be defined for these relationships by means of ASSERT's, if so desired). For example, moving a design object to another cell does not affect its relationships with other design objects. Keeping the relationship matters separate from the organization of design objects in views and cells helps to keep data management simple.

In the work of Katz we also see a shift towards the separation of organization & identification of design objects on the one hand and relationships among design objects on the other hand. In his earlier publications [Kat83, Kat85a], a version object groups multiple design objects (called representation objects) as its different representations. Such a version object serves as an equivalence object, since it groups design objects that are constrained to be equivalent. Thus, equivalence is implied by the way design objects are organized in version objects. In later publications [KBC+87, Kat90], Katz takes the approach that a design object is identified by the triplet <name, version #, type> and proposes three kinds of structural relationships which explicitly relate the uniquely identified design objects.

5.11 SHARING DESIGN DATA ACROSS PROJECTS

In section 5.3 we introduced the notion of *project* to provide logical distribution of design data and design activities. A project is to offer a local environment for performing design activities. In our view, a *library* also is a project, be it that the data contained in it is expected to be referenced quite often and that operation on the data may be subject to more stringent restrictions.

In the previous sections we treated the key data management mechanisms without worrying about the fact that design data is distributed across multiple projects. This was OK since the object type Project was aggregated as an attribute into DesignObject, making each design object a uniquely identified entity. According to the data schema of Figure 5.14, hierarchical, equivalence, and version derivation relationships may relate design objects from different projects. There are no restrictions or explicit constructs modeled in the data schema to handle inter-project relationships separately.

To emphasize the local operation in the project environment, we wish to make references that go across project boundaries more explicit. We adopt a mechanism where relationships within the project context can be established freely, but relationships across project boundaries can occur only if use of the 'foreign' design objects to be referenced has been declared explicitly. This allows explicit selection and administration of objects from (library) projects that are to be used in the course of a particular project.

We adopt a two-staged approach, where first other projects can be declared to be used as libraries for a particular project and then design objects from these libraries can be *imported* into the project. This yields two new object types in the data schema (see Figure 5.18):

```
TYPE Project = ProjectID, Team
TYPE Cell = Name, Project, Designer
TYPE View = Cell, ViewType
TYPE DesignObject = View, Vnumber, Vstatus, PhysLoc,
                                                    Date
TYPE LibraryRef = Project, Lib-Project, LibName
TYPE ImportedDesignObject = [DesignObject], LibraryRef,
                                            Lib-DesignObject
```

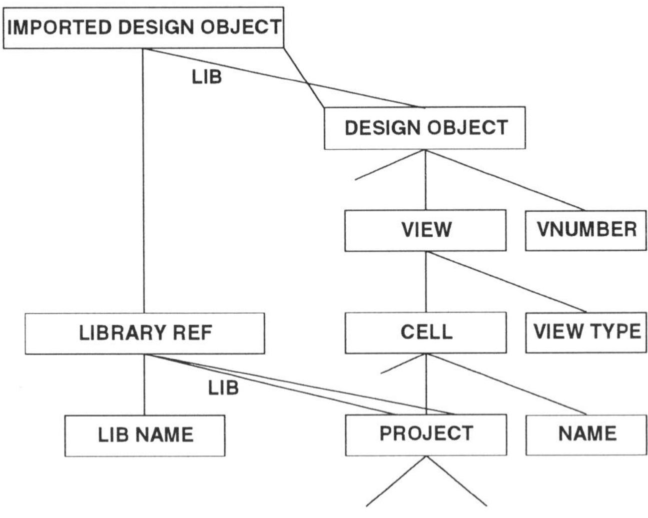

Figure 5.18   Data schema extensions for libraries and imported design objects.

The first new object type is LibraryRef, which represents a reference to a Lib-Project that is to be used by Project, with a local LibName for short reference. The second new object type is ImportedDesignObject, which represents a reference to a design object in a library project. The referenced design object is Lib-DesignObject. Locally an imported design object, though being a reference, is treated as a design object that can be versioned, instantiated, etc. For this purpose ImportedDesignObject has been modeled as a specialization of DesignObject: an ImportedDesignObject IS-A DesignObject. Hence, when a design object is imported from a library project, locally a new (imported) design object is created. The value of the PhysLoc attribute will indicate that for this design object no design data is stored locally. Relatability guarantees that a design object from a library project can be imported only when this project has been declared to be used as a library. Also a library design object can not be removed as long as there are imported design objects in other projects referencing it. In addition, the following constraints are to be enforced:

```
ASSERT ImportedDesignObject ITS ValidProject (TRUE) =
   DesignObject ITS View ITS Cell ITS Project =
                              LibraryRef ITS Project
```

That is, the project in which the imported design object resides must be the project that has the library reference.

```
ASSERT ImportedDesignObject ITS ValidLibrary (TRUE) =
   Lib-DesignObject ITS View ITS Cell ITS Project =
                              LibraryRef ITS Lib-Project
```

That is, the project in which the library design object resides must be the project referenced by the library reference.

References that go across project boundaries have to go via an imported design object that makes the link to a library design object. The hierarchical, equivalence, and version derivation relationships relate only design objects from the same project, which are either local or imported. Such a constraint can be expressed as follows:

```
ASSERT HierarchyRel ITS ValidRel (TRUE) =
   Parent-DesignObject ITS View ITS Cell ITS Project =
       Child-DesignObject ITS View ITS Cell ITS Project
```

Figure 5.19 illustrates the presented concept for importing design objects from library projects.

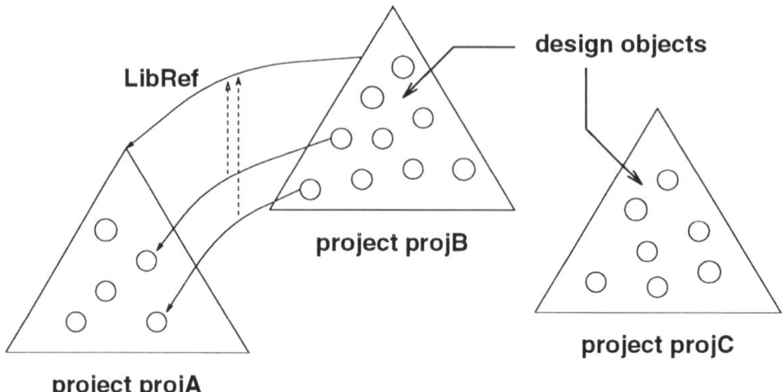

Figure 5.19 Project projB has a library reference to project projA and has imported two design objects from projA.

The library reference from project projB to project projA indicates that projB has declared projA as a library. Project projB has imported two design objects from projA for use in projB. The references to the library objects appear locally as (imported) design objects. These design objects are versions of views, may be instantiated, and may have equivalence relationships, just like local design objects. The primary difference is that the corresponding design data resides with the referenced library objects in project projA.

In summary, the solution presented allows design data to be shared across multiple projects *without copying*. References to library components get administered explicitly and participate as local design objects in the data management procedures employed in the local project environment.

5.12 DESIGN FLOW MANAGEMENT

5.12.1 Introduction

In section 5.5 we modeled the notions of tool run and design transaction in a data schema (see Figure 5.5). Based on this data schema, the framework can maintain and make available information about runs of tools and their accesses on design objects. It provides the basis for a design transaction history facility that informs the end-user about the state of his design, as was also demonstrated by the example queries in section 5.5. A little extra imagination may make visions of additional functionality bubble up in our minds. Consider, for example, the following:

- Runs of tools on design data may invalidate previously derived verification results. The CAD framework should prevent the use of invalidated data in subsequent design steps. The design engineer must be informed about the *validity* of his data and be supported in (re-)obtaining valid verification results.

- Many design steps are highly automatic, i.e. they are performed by design tools that map input data to output data without any user intervention. Execution of such design steps must be *automated*.

- It often occurs that a design engineer is not satisfied with results obtained. The CAD framework should support him in backtracking to a previous starting-point and in efficiently re-executing the design steps from there.

- If the CAD framework is aware of the state of design and is aware of possible ways of transforming this state of design into a "better" state of design, then it can advise the design engineer on tasks to perform next.

Only a very knowledgeable CAD framework can provide such functionality. The recurring theme in the above examples is that the framework exploits knowledge of the design process and the design history to support the design engineer in performing design activities. This includes knowledge of the available tools and the possible ways data can *flow* from one tool to the other. Using this knowledge it can *constrain* the activities of the design engineer by allowing only runs of tools for which valid input data is present, and *support* the design engineer by indicating which tools can or should be (re-)run and possibly run tools automatically for him. This helps to make the design process less error-prone and to improve productivity. We gather these facilities under the term *design flow management*, as they are concerned with the execution of design activities according to a certain design procedure. From an end-user perspective, key design flow management issues are:

- *Design tracking*: keep track of the state of design and the design history.

- *Constraint enforcement*: permit constraints on the design process to be defined and enforced.

- *Guidance and Automation*: support the design engineer in efficiently executing design activities.

We now study in more detail the information needs of a CAD framework with respect to design flow management. The objective is to extend our data schema to represent the object types and relationships relevant to design flow management. We start with the following definition:

Definition 5.4 *A* design flow *is a description of a design process in terms of design activities and temporal, data, and control dependencies between design activities.*

A *design activity* corresponds to a design function which can be performed by a design tool. A single design tool may be able to perform different design activities, for example under control of options passed on the command line. A design flow defines the order in which design activities are to be executed. It specifies *required sequentiality* and *allowed parallelism* for the execution of design activities.

In order to keep track of the state of design, the CAD framework not only has to know the configured design flow, i.e. *what can be done*, but also which design activities have been performed so far, i.e. *what has been done*. We therefore distinguish between configuration information and run-time information:

- The *configuration information* is the pre-defined design flow. It is a template defining placeholders for actual data. It is defined before the actual design process is started, and modified only when the design procedure is adjusted. For example, this may occur when a new (version of a) tool is made available or when a tool is no longer to be used. Thus, the configuration information is relatively stable.

- The *run-time information* is updated continuously in the course of the design process, in order to correctly administer the state of design. It includes information on tool runs, the activities performed during tool runs and the data involved. The run-time information 'colors' the template design flow by filling the placeholders with actual data items consumed and produced during actual tool runs. This information can be used to inform the design engineer about the state of his design, to check the validity of data, to check whether input conditions for tools are satisfied with respect to the pre-defined design flow, etc. That is, to provide the advanced functionality imagined above.

We note that the data schema derived in section 5.5, containing the object types Tool, ToolRun and DesignTransaction, already represented both kinds of information. The information on tools can be considered configuration information, and the information on tool runs and design transactions can be considered run-time information. However, this data schema does not represent information on the individual design activities that may be performed by the tools and the dependencies between design activities. It must be extended to provide a basis for design flow management.

5.12.2 Design Flow Configuration

There has been some discussion in the framework community on the use of configured design flows. Design tracking can be performed without having a pre-defined design flow. In fact, the design transaction history facility presented in section 5.5 is a simple design tracker. A prominent example is the VOV system of Casotto et. al. [CNSVl90], which captures the design history without relying on a configured design flow. However, there are several advantages to the use of configured design flows:

- It permits *constraint enforcement*: constraints on the design process can be defined in the design flow to be enforced by the design flow management system. Rumsey et. al. [RF92] mention that flow-based approaches are not suited to design styles that require a lot of freedom, e.g. ASIC design styles requiring what-if experimentation. We do not share the view that the use of configured design flows in itself imposes restrictions on the design process. A proper design flow management system allows tool sequences to be configured according to the user's wishes. It is up to the flow configurator to allow all tool sequences that are technically possible, to just prevent mis-use of data, or to be more restrictive and have particular tool sequences enforced. That is, he can choose the favorite position in the flow enforcement spectrum.

- It helps to provide *guidance* and *automation*. Since the design flow management system can learn from the configured design flow what may be done to the design given its current state, it can *guide* the design engineer by informing him about enabled tools for particular data, and possibly run tools automatically. Since the design flow management system can also infer from the configured design flow how a particular state can be reached from its current state, it can support *goal-directed* automation. Configurable design flows can be used for design planning. That is, tools and partial design flows can be selected explicitly before they are actually used. In [SBD93], Sutton et. al. introduce the concept of dynamically defined flows - partial flows that are constructed, on demand, by a design engineer subject to the rules contained in a configured task schema (a task schema corresponds to a design flow, and defines all possible partial flows).

- It permits the construction of a flow-based *user interface* based on the principle of *flow coloring* [tBvdWB93]. Such a user interface informs the end-user in an attractive way about the state of design by 'coloring' the configured design flow with data items and tool states. It permits the design history to be queried conveniently, using the configured design flow as a graphical query template. See also section 6.9. Kleinfeldt et. al. [KGMB94] report a trend toward graphical flow definition.

We stress that flexibility in flow configuration is key. The design flow must be easily configurable and must be allowed to evolve, for example, when a new tool is to be used.

A design flow is defined in terms of the constructs provided by a *design flow model*. According to Definition 5.4, a design flow model must offer constructs to describe design activities and dependencies between design activities. An attractive paradigm for a design flow model is the data flow paradigm. The RoadMap [vdHT90] and Nelsis [tBBvdW91] design flow models are both based on the data flow paradigm. We present a similar design flow model and formally define the relevant object types and relationships in a data schema. Related references are [vdHtBB$^+$91], [BtBvdW92] and [tBvdWB93].

A central notion in the design flow model is the *flow graph*. A flow graph corresponds to a functional unit that maps input data to output data. A flow graph communicates with its environment through *ports*. A port is either an input port or an output port and handles data of a particular *data type*. Access via a port is either required or optional. Design activities, i.e. design functions which are performed by design tools, can be modeled by flow graphs. The defined flow graphs then represent the possible functional transitions on the state of design. As we said, a single design tool may be able to perform different design activities. Hence, multiple flow graphs may be defined for a single tool. *Tool characterization* is the definition of flow graphs for the design activities that may be performed by a tool. It is the first step towards incorporation of a selected tool into a design flow. The data types of the ports of the flow graphs relate to the view types used for the classification of design objects. For reasons of flexibility we permit multiple data types to relate to the same view type. For example, for the view type 'circuit' there can be a data type 'netlist' and a data type 'expanded netlist'. A design object of view type 'circuit' may then contain both data of data type 'netlist' and data of data type 'expanded netlist'.

In order to describe dependencies between design activities we introduce the notion of *channel*. Channels can be used to connect ports of flow graphs. Data can *flow* from an output port of one flow graph to an input port of another flow graph only if a channel connects these two ports. That is, channels define producer - consumer relationships between the tool functions of the configured set of tools. A channel may connect multiple output ports to multiple input ports. All ports connected by a channel must be of the same data type. Multiple channels may be linked to a port. Note that after tool characterization for a set of selected tools, channels may be generated automatically by having all ports of the same data type connected by a channel. The result is the design flow that permits all tool sequences that are technically possible for the set of selected tools. Such automatic channel generation may help cut setup time upon design flow configuration.

Ports and channels can be used to express logical conditions for the execution of design activities. Multiple input ports of type 'required' for a design activity implies a logical AND, since valid data must be present for all these input ports. A channel that permits multiple producer design activities to 'feed' an input port of a consumer design activity implies a logical OR. The design flow model thus permits conditions to be expressed as AND's of OR's.

We permit flow graphs to be defined *hierarchically*: compound flow graphs may be composed from more primitive flow graphs. This permits the definition of well-structured design flows, as details can be hidden inside compound flow graphs that represent higher level design tasks. The individual tool functions are the basic functional units, that is, the design activities are the leafs in the flow hierarchy. Compound flow graphs, like leaf flow graphs, communicate with their environment through ports. The ports connected by a channel in a compound flow graph are either port of the compound flow graph or port of a component flow graph (port instance).

The key constructs of the design flow model are illustrated by the example compound flow graph in Figure 5.20. The example also shows how logical conditions for the execution of design activities can be expressed by combining ports and channels.

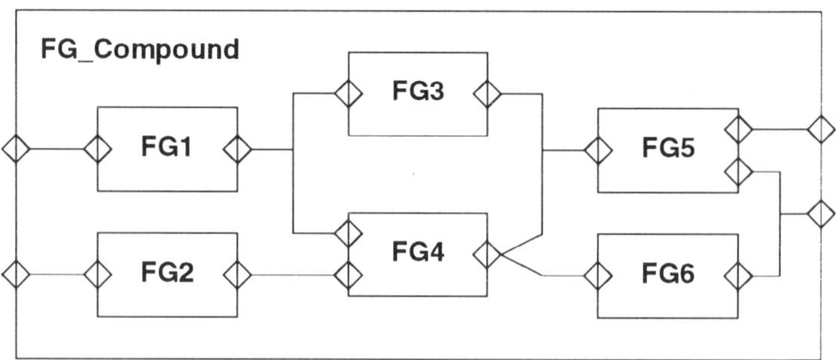

Figure 5.20 Example compound flow graph, containing instances of other flow graphs, with ports connected by channels. Flow graphs are represented by rectangles, ports by diamonds, and channels by lines connecting diamonds.

The example in Figure 5.20 shows that flow graph FG1 can feed both FG3 and FG4, where FG4 also needs data produced by FG2 (logical AND). FG3 and FG4 can both feed FG5 (logical OR), but only FG4 can feed FG6 (via a separate channel), even though FG3 produces data of the same data type. The ability to link multiple channels to a port can be useful in modeling real-life design procedures. For example, both a layout-to-circuit extractor (FG3) and a schematic editor (FG4) produce a netlist that can be simulated (FG5), but only the netlist from the schematics can serve as input for a place & route package (FG6). Notice again the data classification capabilities of the flow paradigm, with reference to the discussion about view types in section 5.6. A design object is classified not only by its view type, but also by the design activity that produced it.

Now that we have recognized relevant concepts for design flow configuration, we want to represent them with the OTO-D data modeling technique. This brings us to the following (backbone) data schema (see also Figure 5.21):

```
TYPE FlowGraph = FGName
TYPE Activity = Tool, [FlowGraph]
TYPE DataType = ViewType, DTName
TYPE Channel = FlowGraph
TYPE Port = Channel, DataType, PType
TYPE FlowHierarchy = Parent-FlowGraph, Child-FlowGraph
TYPE Link = FlowHierarchy, Channel, Port
```

The concept of flow graph is represented by the object type *FlowGraph*. A flow graph is identified by a name. Design activities (i.e. tool functions) are represented by the object type *Activity*, having the attribute *Tool*. The specialization relationship with FlowGraph tells that an Activity IS-A Flow-Graph. A data type has a name and relates to a view type. Multiple data types may relate to the same view type. A channel belongs to a flow graph. A port relates to a channel of a flow graph, has a data type and a port type (required / optional, input / output). Channels in leaf flow graphs are used to cluster ports via which different types of data belonging to the same design object are to be accessed. Such a *port set* construct is required since we permit multiple data types to relate to the same view type. This permits a design object to contain different types of data which are to be accessed via different ports. The object type *FlowHierarchy* represents the hierarchical relationships between parent (compound) and child (component) flow graphs. The object type Link represents links of ports of component flow graphs (i.e. port instances) to channels in compound flow graphs. A port may be linked to multiple channels in a compound flow graph, as illustrated by the output port of FG4 in Figure 5.20.

In section 5.5 we touched the subject of project-specific tool selections and promised an elegant solution in the context of design flow management. As we have seen in this section, a hierarchical design flow can be defined for a set of selected tools. This design flow is identified by its root flow graph. We refine the definition of *Project* and give it the attribute *FlowGraph*:

```
TYPE Project = ProjectID, Team, FlowGraph
```

This permits each project to have a selected design flow and thereby a set of selected tools.

5.12.3 Run-Time Operation

In the course of the design process, the design flow management system must capture and administer run-time information to keep track of the state of design and the design history. This is referred to as *design tracking*. Constraint enforcement and guidance and automation services depend on proper design tracking. Publications focusing on design tracking are [CNSVl90, VMT92, BtBvdW92].

According to our design flow model, the leaf flow graphs in a hierarchical flow definition model the design activities that may be performed by the design tools, and represent the possible functional transitions on the state of design. In the course of the design process, the design tracking facility must identify and administer the design activities that are performed and the data involved. We permit a tool to perform any of its activities any number of times during a single tool run. This is to support tools that may perform several activities during a single run, possibly in response to interaction with the end-user. Because of the ability to describe and track the design process at the level of design activities rather than at the level of tool runs, we speak of *fine-grain* design flow management.

As design activities are performed, the configured constraints must be enforced. An activity is allowed to access data only if it has an input port of the proper data type which is connected by a channel to the output port of the design activity via which this data was produced. An activity is allowed to produce data only if it has an output port of the proper data type. An activity is *executable* with respect to a set of design objects if for all required input ports data may be accessed. An activity has *completed successfully* if it has produced data for all its required output ports.

We represent the design flow management run-time information in our data schema (see also Figure 5.21):

```
TYPE ToolRun = Tool, Project, Used-Opts, Start-Date,
                                                  Designer
TYPE RunningDesTrans = DesignObject, ToolRun, Date,
                                                  AccMode
TYPE ActivityRun = ToolRun, Activity, ARMode
TYPE DesignTransaction = DesignObject, ActivityRun,
                         Channel, Date, AccMode, ComplMode
```

Tool runs, either in progress or completed, are administered via the object type *ToolRun*. While a tool run is in progress it can hold access rights on design objects via its running design transactions. These are represented by the object type *RunningDesTrans*. In the course of a run, a tool may perform multiple runs of configured activities. The CAD framework identifies the activities that the tool is performing and decides on their validity. The *activity runs* get administered via the object type *ActivityRun*. When a design transaction completes it is administered via the object type *DesignTransaction*, with a reference to the corresponding activity run and the internal channel (= port set) of the activity via which the design object was accessed. On this occasion the running design transaction is deleted, thereby withdrawing the access rights from the tool run.

Compared to the data schema presented in Figure 5.5, the definition of the object type DesignTransaction has been refined, as it now refers to an individual activity run rather than a tool run. The CAD framework administers which individual design objects are involved in the activity runs: which objects are used in which functional transitions to produce which other objects. In this way the complete history is retained and validity of data can be judged. Moreover, the CAD framework can identify activities for which all input data is available and inform the design engineer, or run these activities automatically for him. Since the complete history is retained for all objects still alive, backtracking to previous design states is supported.

Figure 5.21 depicts the diagram for the backbone of the defined data schema. The data schema consists of three main parts, as has been indicated by the dotted lines in Figure 5.21:

- The *data management part* defines the structure of the information that is maintained by the data management services. The central object type is DesignObject. Most of the object types defined in the previous sections are located in this part.

- The *flow configuration part* defines the structure of the configuration information.

- The *run-time part* defines the structure of the run-time information that is maintained to keep track of the state of design.

The single data schema, which captures the definitions of the different kinds of information, provides the basis for a CAD framework that manages these kinds of information in a unified meta data database.

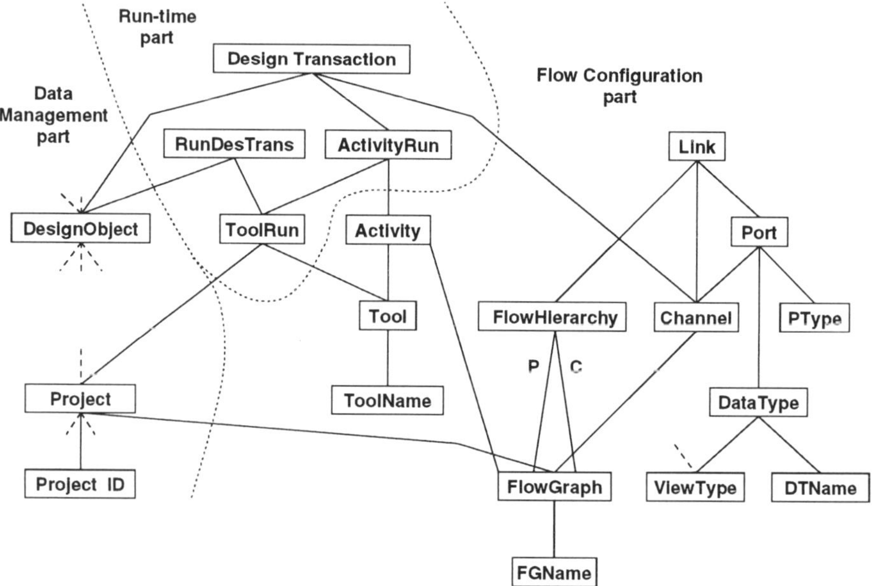

Figure 5.21 Backbone of the data schema for design flow management.

5.13 CONCLUSION

In this chapter we have defined the information architecture of a CAD framework. It describes at a domain neutral level the logical organization of the design environment in terms of object types and their relationships, and is a key view of the framework architecture.

The OTO-D semantic data model, presented in the previous chapter, provided us the formal means for deriving and representing the information architecture in the form of a data schema. The data schema was defined incrementally by studying different aspects of the management of design data and the design process. In a step-by-step procedure, object types were defined and refined. The resulting data schema is:

```
TYPE FlowGraph = FGName
TYPE Project = ProjectID, Team, FlowGraph
TYPE Cell = Name, Project, Designer
TYPE View = Cell, ViewType
TYPE DesignObject = View, Vnumber, Vstatus, PhysLoc,
                                                 Date
TYPE Tool = ToolName, Path, Possible-Opts
TYPE Activity = Tool, [FlowGraph]
TYPE DataType = ViewType, DTName
TYPE Channel = FlowGraph
TYPE Port = Channel, DataType, PType
TYPE FlowHierarchy = Parent-FlowGraph, Child-FlowGraph
TYPE Link = FlowHierarchy, Channel, Port
TYPE ToolRun = Tool, Project, Used-Opts, Start-Date,
                                                 Designer
TYPE RunningDesTrans = DesignObject, ToolRun, Date,
                                                 AccMode
TYPE ActivityRun = ToolRun, Activity, ARMode
TYPE DesignTransaction = DesignObject, ActivityRun,
                         Channel, Date, AccMode, ComplMode
TYPE LibraryRef = Project, Lib-Project, LibName
TYPE ImportedDesignObject = [DesignObject], LibraryRef,
                                                 Lib-DesignObject
```

```
TYPE VersionDerivationRel = Original-DesignObject,
                                    Derived-DesignObject
TYPE HierarchyRel = Parent-DesignObject,
            Child-DesignObject, InstName, Constructor
TYPE EquivalenceRel = Source-DesignObject,
                Target-DesignObject, ToolRun, Class
```

In addition to these definitions of object types and attribute relationships, extra static constraints have been defined as assertions. These are not repeated here. The backbone of the data schema is depicted in Figure 5.22 (the flow configuration part has been partly omitted).

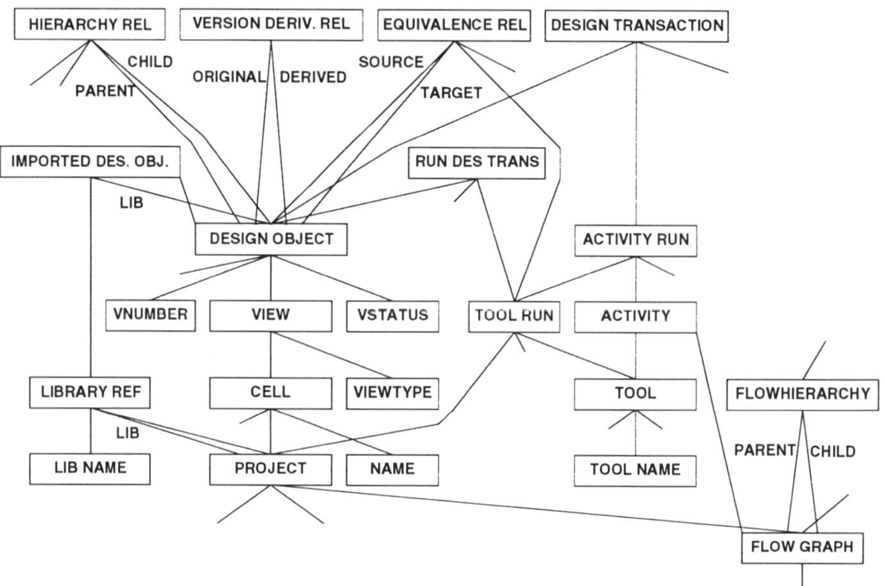

Figure 5.22   Backbone of the CAD framework data schema.

The data schema is relatively simple and comprehensible, yet represents the information structure related to the key aspects of data and design management in a CAD framework. The central object type is DesignObject, and many other object types relate to it. This is a direct reflection of our conclusion in section 3.3 that the framework manages design data at the coarse-grain level of design objects. Based on the data schema, the CAD framework organizes design objects across views, cells and projects, maintains relationships among design objects, and keeps track of their state. The information-rich pool of meta data that becomes available while operating, can be used to keep the design engineer well-informed about the structure and status of the design and assist him in selecting the design activity to perform next.

As we said before, data modeling helps in achieving understanding of the information needs of an organization and by extension the way in which an organization functions. In the definition of the information architecture many principal choices have been made, and as such it defines the context for the definition of the detailed functionality of the individual framework services. Actual definition of the detailed functionality may yield some data schema refinements, but this will not affect the backbone of the data schema (provided no serious defects are encountered). An attractive property of the OTO-D data modeling technique is that the resulting aggregation hierarchies offer different levels at which additional attributes can be attached. Openness of the CAD framework was secured by avoiding incorporation of features of a particular tool set or design representation in the definition of the data schema.

Our work relates to the work of e.g. Katz [KAC86, KBC+87] and Batory [BK85] in that it investigates the major organizing principles employed by the data and design management facility of a CAD framework. What sets our work apart is that we formally represent the structural semantics of these principles at the *type* level by means of a *single* data schema. Katz presents multiple example pictures at the instance level, showing possible occurrences of objects and relationships. In contrast we have focused on representing the *invariants* of a design environment. The data schema makes the relationships between all object types explicit and, as it has been constructed with well-defined abstraction primitives, has an unambiguous interpretation.

Recently, we see more examples of the application of data modeling techniques in the area of CAD frameworks. Some publications presenting a data modeling approach to data and design management for CAD frameworks are [vdWvL88], [SZ89], [vdHT90] and [BK92].

CFI has also recognized the importance of data modeling to their standardization activities. They apply data modeling techniques in the definition of standards for handling domain-specific types of data, as well as for defining views on an overall framework architecture [CFI93]. An interesting reference in this context is [vdH91], which strongly motivates the application of data modeling techniques for CFI's architecture-related activities.

The power and relevance of the application of data modeling techniques in the definition of CAD framework architectures is demonstrated by the work presented in [vdHtBB+91]. This report presents a comparison between two design flow paradigms, namely the RoadMap paradigm developed by van den Hamer and Treffers from Philips Research Laboratories, as presented in [vdHT90], and the Nelsis paradigm as presented in [tBBvdW91]. The comparison was performed by actually inspecting the data schemas of both systems, seeking for correspondences and differences in information structures. The striking conclusion from this work is that starting from the inspection of the data schemas a detailed insight was obtained in the correspondences and differences in capabilities of both systems.

6

THE COMPONENT ARCHITECTURE

6.1 INTRODUCTION

In this chapter we present the component architecture of the CAD framework. This architectural view identifies the individual framework components and the dependencies between them. It provides additional detail on the logical structure of the framework.

In chapter 3 we looked into some global issues and derived a global framework model (see again Figure 3.4), from which we now proceed. We summarize the outcome of chapter 3 as follows:

- The CAD framework consists of a *framework kernel* and *framework tools*. All interaction with the end-user is through tools.

- The framework kernel keeps track of the *state of design* and controls all access to the data.

- We logically distinguish between *meta data* and 'raw' design data contained in *design objects*. Meta data refers to design objects as logical units.

- Interaction between tools and framework kernel takes place via the *tool - framework interface*.

In chapter 5 we derived the information architecture of the CAD framework. As stated in section 3.5, this is considered a *user-view* of the framework. The data schema presented in Figure 5.22 represents the information structure related to the key aspects of data and design management. From this data schema we learn which meta data is to be maintained by the framework about the design objects.

Our next goal is to obtain a more detailed understanding of the internal structure of the framework. For this purpose we will define the component architecture. We will identify principal framework functions and allocate them to mutually ordered components. The component architecture is considered a *developer-view* of the framework. Note that components are intended only to represent the logical divisions of framework functions and do not imply implementation as a single unit. In the following sections we will identify logical framework components, and decompose them into smaller components. Functionality will be allocated to components, and interfaces will be defined.

The component architectures derived in the following sections will be represented by diagrams built of boxes and arrows connecting boxes. A box represents a *logical component*, which provides a group of *functions*. Components make their functions available to other (client) framework components or design tools by exporting them via their *interface*. The interfaces will be represented in the diagrams by shaded areas. The arrows connecting the boxes represent *calling dependencies* between the components: which component causes which other component to perform a function. The principal interface functions will be presented textually.

6.2 FRAMEWORK KERNEL AND FRAMEWORK TOOLS

We use the term *framework service* to denote a logical group of functions, which offers a particular facility to the design tools and/or the end-user. Some example framework services are (see also the principal requirements in section 2.3):

- Concurrency control.

- Access control.

- Version management.

- Support for hierarchical multi-view design.

- Design flow management.

These services are considered to be common among CAD applications, and therefore are to be provided at the level of the underlying infrastructure, the CAD framework.

The only framework components identified so far are the *framework kernel* and the *framework tools*. A key aspect of our global framework model (Figure 3.4) is the presumed run-time interaction between design tools and the framework kernel. That is, while they are running, design tools may transfer control to the framework kernel in order to have framework functions activated. For this purpose, the framework kernel makes a set of possible requests available through the tool-framework interface. There is no direct interaction between design tools and framework tools.

Framework functions are located either in a framework tool or in the framework kernel. We define the following discriminating principle, which can be applied upon the allocation of framework functions to components.

Principle 6.1 *If a framework function is to be activated upon the execution of a request issued by a tool to the framework kernel, then it must be located in the framework kernel.*

Functions intended to be activated solely by the end-user may be located in a framework tool. A function that is located in a framework tool may use lower functions located in the framework kernel, via the tool - framework interface.

For example, version handling functions that have to be active when design transactions are performed by design tools, are part of the kernel. A graphical version browser for the end-user is located in a framework tool. This version browser may call kernel functions to retrieve information about the available versions and their relationships.

In order to further qualify the kernel framework functions, we must be more specific about the requests that tools are allowed to issue to the framework kernel. We take a look at the kind of interaction that design tools may have with the framework kernel when performing their specific design tasks. Design tools operate on design objects and their relationships. Do we allow a design tool to decide *at run-time*:

- On which design objects and relationships to operate?

- What to do with these design objects and relationships?

For maximum *openness* and *flexibility*, we answer both questions with a loud and clear "yes, we do". The actual behavior of a design tool (from a framework perspective) may be influenced by:

- Command line arguments.

- User interaction.

- Contents of command files.

- Characteristics of the design.

Hence, we consider it absolutely essential to allow design tools to decide *at run-time* what to do with which design objects and relationships. As a consequence (principle 6.1), all functions that are involved in controlling and administering the accesses of design tools to design objects and their relationships must be located in the kernel. This includes the major data and design management functions. We therefore conclude that the framework kernel really is the heart of the CAD framework.

As a consequence of this design choice, the CAD framework can handle interactive editors that may perform multiple edit operations during a single run, with design objects being selected interactively by the design engineer.

It is also able to keep track of tools that seek their way through the multi-view design hierarchy, taking smart decisions on which operation to perform where.

Although our choice to allow a large degree of run-time interaction may appear rather obvious, history has shown different approaches. Some (prototype) frameworks were aimed exclusively at a tool integration technique known as *encapsulation* (section 6.8). An example are the early versions of the Jessi-Common-Framework [JCF91b, LJ92]. A major component of these framework versions is the *desktop*, a graphical framework tool from which all tool runs have to be launched. The major data and design management functions are located in the desktop, in particular the functions to check and update the design flow status. Upon a tool run, the desktop puts the design files in place, activates the tool, and collects the resulting files. The new state of design is inferred by inspection of the files produced by the tool run. There is no interface that allows tools to interact at run-time with a framework kernel. As a result, the framework can handle only tools for which the appropriate files can be determined beforehand.

In contrast, we locate the major data and design management functions in the framework kernel, and allow design tools to interact at run-time with this kernel. This approach was also advocated in [tBBvdW91] and [BtBvdW92] in the context of design flow management. Key advantages are:

- Openness to a wide range of tools.

- Support of the full tool integration spectrum, from encapsulation to tight integration (see section 6.8).

- Accurate administration of the state of design.

- Early updates of the administration, i.e. while the interaction occurs.

- Potentially the most efficient approach, as tools inform the kernel about their accesses. No post-mortem inference of the design state is required.

- Tools are not required to be invoked from a special command shell.

6.3 THE FRAMEWORK KERNEL

We direct our attention to the framework kernel, in order to decompose it logically into smaller components. In this section we will make an initial decomposition, which will reveal more detailed issues that are to be addressed in the following sections.

In chapter 3 we characterized the framework kernel as a transaction processing system, through which tools can perform transactions on the state of design. One key aspect to this is the *storage* of all information pertaining to the state of design. The framework kernel has to offer permanence of the results of committed transactions. This relates to the framework's role of design database, as discussed in chapter 2. We therefore identify a *Data Handling* component, which provides facilities for reliable persistent storage of meta data and design data. These are *generic* facilities; the Data Handling component has no built-in knowledge of the semantics of the data that it handles. It offers functions for creating, storing, retrieving and deleting data.

In section 3.3 we stated that the design database provided by the CAD framework is to be a generic facility. Support for specific design representation formats is not a principal framework responsibility, but may be offered through *configuration* of the generic Data Handler. We do *not* include an explicit framework component for this type of functionality, as is e.g. done by CFI [CFI93] (see again Figure 2.4). Below we will demonstrate the openness of the framework architecture to support for specific design representation formats.

The framework kernel is much more than a 'dumb' storage component. In the previous section we concluded that a variety of data and design management functions are located in the framework kernel. They have to take care of issues such as access control, data organization, and design process control. For example, when a design tool performs a design transaction on a design object, these functions may check access rights, check and update the object's design flow status, and handle versioning issues. We allocate these functions to a component called *Data and Design Management Kernel*. We will also refer to this component as the *DDM Kernel*. The DDM Kernel gets informed about all transactions performed on the system. It keeps track of all design activities, and administers the relevant meta data.

We position the DDM Kernel on top of the Data Handling component. As opposed to the Data Handling component, the DDM Kernel has built-in knowledge of the semantics of the meta data. The DDM Kernel uses the Data Handling component as a storage component, for example, to hold its meta data administration. The Data Handling component calls no functions from the DDM Kernel. Thus, we structure the framework kernel as a 'brain' component, offering *specific* data and design management functions on *specific* types of meta data, on top of a 'muscle' component offering *generic* storage facilities. The advantage is increased modularity, which facilitates framework evolution. The generic Data Handling component can easily adjust when new data and design management functions are to operate on new types of data.

The basis of the complete environment is the platform, i.e. the hardware and the operating system software, on which the framework and tools are to run. We use the term *system environment* to denote the functions typically associated with an operating system. These include facilities to spawn and run processes, I/O handling, a hierarchical file system, protection mechanisms, interprocess communication (IPC) facilities, and network services.

On top of the system environment, a whole range of general-purpose facilities may be available for the benefit of all framework components and the design tools. These may include window-based graphics (e.g. the X Window System, OSF/Motif), portability services, advanced network services, a remote procedure call (RPC) mechanism, error handling, license management, etc. A trend we have seen over the last years, is that more and more of these general-purpose facilities are migrating to the system environment. They are becoming standard facilities, usually offered by hardware vendors or specialized software vendors. This trend is expected to continue.

We capture the functions offered by the system environment *and* the general-purpose facilities on top of it, in a single base component termed *System Environment and Common Basic Services*. The functions provided by this component can be used by all other components, including the tools.

We represent the identified kernel components with the calling dependencies in Figure 6.1.

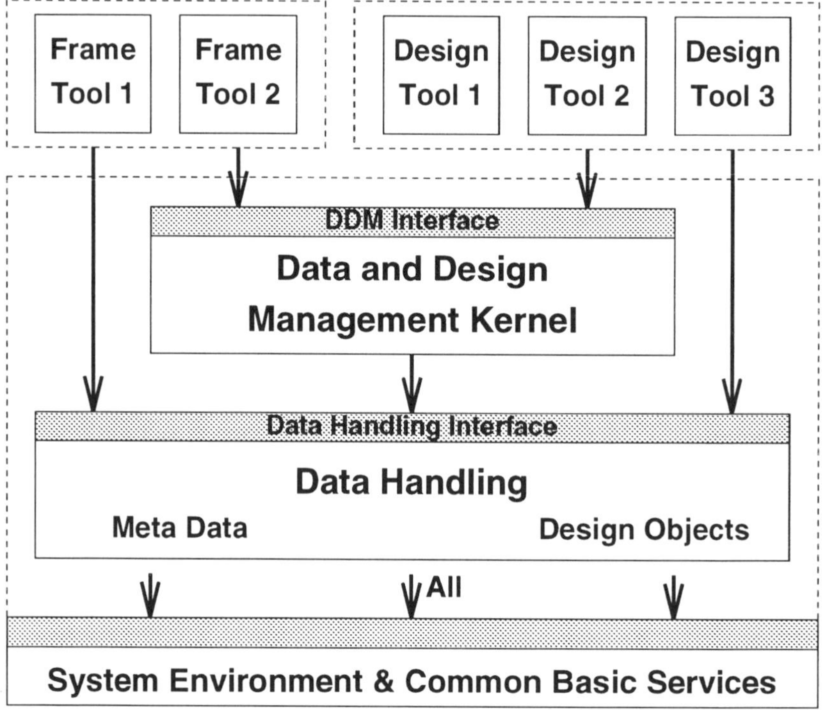

Figure 6.1 The framework kernel is logically structured as a general base component with a Data Handling component and a Data and Design Management Kernel stacked on top of it.

The System Environment and Common Basic Services component is drawn at the bottom of Figure 6.1. It offers a great variety of functions to all other components. This component will not be further detailed.

The Data Handling component provides a facility for reliable persistent storage of meta data and design data. Higher level components interact with the Data Handling component via the *Data Handling Interface.* The internal structure, functionality and main characteristics of the Data Handling component will be further detailed in the next section. In particular we will address the partitioning of this component into smaller components, and the openness to a variety of storage regimes for the 'raw' design data.

The DDM Kernel component is built on top of the Data Handling component. Tools interact with the DDM Kernel via the *Data and Design Management Interface.* We will refer to this interface as the *DDM Interface.* In section 6.7 we will discuss the DDM Kernel in more detail.

We stated that all interaction between tools and framework kernel takes place via the *tool - framework interface.* Now that the framework kernel gets decomposed into multiple smaller components, which have interfaces that may be used by the tools, we must think of the tool - framework interface as a *set of interfaces.* Both framework tools and design tools can activate data and design management functions via the DDM Interface. This interface hides many of the underlying details from the tools. It makes a pre-defined set of transaction types available. One of the transaction types will be the *design transaction,* as defined in section 3.4, which allows design tools to operate on design objects. We elaborate on the DDM Interface in section 6.8. Framework tools also interact directly with the Data Handler to access meta data. Framework functions located in a framework tool use the same interface for meta data access as the functions located in the DDM Kernel. Design tools interact directly with the Data Handler to access design data. This is in agreement with the statement made in section 3.3, that framework intervention in handling detailed design data is minimized. We note that this design data access is allowed only if access rights have been granted by the DDM Kernel. In the following sections we will further detail the interfaces and define the constraints for their usage.

6.4   THE DATA HANDLING COMPONENT

6.4.1   Key Data Handling Issues

The Data Handling component provides a facility for reliable persistent storage of all data handled by the CAD framework. It offers functions for creating, storing, retrieving and deleting data. Below we list some key issues for this component. Most of these issues relate to our principal requirements presented in section 2.3.

- *Support appropriate data types and access methods.*

  To avoid an 'impedance mismatch' with the higher level components, the Data Handling component must provide an appropriate set of primitive data types, appropriate constructs for relating data, as well as appropriate methods for accessing data. In other words, the *data model* implemented by the Data Handling component must match the needs of the variety of applications built on top of it. The ability to effectively handle a great variety of data is an important aspect of *openness*.

- *Performance.*

  The performance of the Data Handling component is a critical factor in achieving overall efficiency.

- *Access control.*

  For purposes of security, the Data Handler has to check permissions of users to access data. Note that this has to be balanced with access control mechanisms implemented at the level of the DDM Kernel, in order to prevent an overkill of checking procedures.

- *Consistency and integrity.*

  The Data Handling component must guarantee correctness of the data with respect to pre-defined consistency and integrity constraints.

- *Concurrency and recovery: Transactions.*

  A transaction facility must allow the data to be transformed atomically from one consistent state to another consistent state. See again section 3.2.

- *Logical distribution.*

  The principal requirement for logical distribution (section 2.3) is likely to have an impact on the Data Handler. In the data handling context we define logical distribution as follows: A database system is logically distributed if the user sees and interacts with the system as if there are multiple databases, each individual database having its own data schema [Lyn84].

■ *Physical distribution.*

The Data Handler must operate in a distributed computing environment. It must allow data to be accessed from any location in this environment. Note that the required capabilities of the Data Handling component with respect to physical distribution largely depend on the overall implementation architecture. This is covered in the next chapter.

While taking these issues into account, we will present some key design choices for the Data Handling component and reflect them in the component architecture.

6.4.2 Separate Meta Data and Design Data Handling

The Data Handling component as depicted in Figure 6.1, provides storage facilities for meta data as well as design data contained in design objects. What are the characteristics of these types of data and their access methods? We know that the design objects contain localized collections of related data, but their granularity may vary significantly. The overall volume of the design data that is to be handled in the course of a design project can be huge. Access characteristics may vary significantly, depending on the application. Some design tools require massive bulk access, while others require traversal-based access methods to navigate through the data. Design objects can be involved in long design transactions, thereby prohibiting conflicting accesses by other users to the related design data.

In contrast, the meta data typically is *small* in size when compared to the volume of the corresponding 'raw' design data. *Queries* for small amounts of meta data are issued *frequently* by the DDM Kernel and the framework tools. The meta data must have a *high availability* to allow the DDM Kernel and the framework tools to consult and update the administered state of design without significant delays.

In order to get more specific about the Data Handling component, we first wonder whether it has to be one indivisible logical component. Does the architecture allow:

■　Separate data handling components for meta data and 'raw' design data?

And if so, can we define an interface to allow:

■　Incorporation of alternative data handling components for 'raw' design data?

The advantage of having separate components for Meta Data Handling and Design Data Handling is significant. Each component can be geared towards the characteristics of the data it has to handle. There is no need to build a comprehensive facility that tries to satisfy the disparate needs of meta data and design data handling. In particular we gain the flexibility of having a light-weight component which is optimized for efficient handling of the meta data. The advantage of the ability to incorporate alternative design data handling components is increased flexibility and openness.

In [MGSW89] the importance of a single data handling system is advocated. The key issue that we face here is *consistency*. The risk of employing separate data handling components is that the consistency between meta data and 'raw' design data may be corrupted. We recall the statement made in section 3.3, that the meta data administration should at all times correctly reflect the state of the actual design data as contained in the design objects. The architecture may allow separate data handling components for the meta data and the 'raw' design data if and only if the consistency between the two types of data can be guaranteed under all circumstances.

In order to maintain consistency, we must somehow synchronize the execution of the operations performed on both data handlers. It is important to note, however, that we do *not* have to solve the general problem of maintaining consistency among two arbitrary types of data handlers. We can take advantage of some of the design choices we have presented so far. The key to providing *meta data - design data consistency* lies in the fact that operations on design objects have to be initiated and terminated via interaction with the DDM Kernel. This puts the DDM Kernel in control at critical stages of the design data manipulation process performed by design tools. At these stages the DDM Kernel can synchronize the execution of design data operations with the corresponding administrative operations on the meta data.

Our approach to maintaining meta data - design data consistency is based on the following principle:

Principle 6.2 *At any moment, the Design Data Handling component contains* at least *the design data that corresponds to the meta data registered in the Meta Data Handling component.* Validity *of design data depends on the meta data.*

Principle 6.2 implies that the meta data is 'in charge' of the design data. When the meta data signals presence of particular design data, this design data must be present. When the meta data signals absence of particular design data, this design data should not be present or, if present, is considered invalid. If we adhere to principle 6.2, the system always contains the design data corresponding to the administered meta data. When a tool requires access to a design object, we can always determine which design data to take for the object, based on the actual state of the meta data. We show how principle 6.2 can be adhered to.

We require the Meta Data Handler to offer a *transaction facility*, which permits operations on the meta data to be performed atomically. We call a sequence of meta data operations that is performed atomically a *meta data transaction*. We require the Design Data Handler to offer the property of *permanence*: once updates have been committed, the results will not be lost.

We distinguish three types of operations on design objects: *creation*, *update*, and *removal*. We adopt the following procedures for the three types of operations:

- *Creation* of a new design object:

  This operation is performed by means of a design transaction (Check-Out - CheckIn sequence). Upon CheckIn, the new design data must be committed before committing the design transaction in the meta data administration. The meta data decides on the validity of the new design data.

- *Update* of an existing design object:

  This operation is performed by means of a design transaction (CheckOut - CheckIn sequence). The DDM Kernel decides whether the new design object becomes a new version or that it is to overwrite the existing design object. Existing design data may not be overwritten in the course of the design transaction. As explained in section 3.4, the tool operates on a (virtual) copy of the design object. Upon CheckIn, the new design data must be committed before committing the design transaction in the meta data administration. The design object temporarily has a 'double state'. The meta data decides which state is the valid one. Obsolete design data can be removed only after the administration of the design transaction in the meta data has been committed.

- *Removal* of an existing design object:

  For this purpose, we introduce the RemoveDesignObject operation in the DDM Kernel. This operation first administers the removal of the design object in the meta data, and only then removes the corresponding design data. The design data is effectively invalidated by the update performed on the meta data.

The essence of the above procedures is that we carefully *sequence* the operations on both data handlers. Upon a meta data transaction it always holds that the design data of the old state *and* the design data of the new state are both present. The meta data transaction atomically transfers the validity from the old state to the new state. As a result, the correct state of the design data, as administered in the meta data, can always be obtained. Note that it is a matter of proper implementation to ensure that all intermediate stages in the sequencing of operations are well-defined and identifiable. Corresponding recovery procedures must take care that uncompleted design data operations are completed correctly, in accordance to the design state administered in the meta data. Since the DDM Kernel is in control, this can be handled correctly.

On the basis of the above procedures, principle 6.2 is adhered to, and meta data - design data consistency can be guaranteed. As a result, we can refine our component architecture to allow separate data handling components for meta data and design data contained in design objects. This is depicted in Figure 6.2.

Figure 6.2 shows that the Data Handler has been split into a Meta Data Handling component and a Design Data Handling component. The tool - framework interface has not been drawn explicitly. We can now be more specific about the numbered calling dependencies in Figure 6.2. The arrow tagged with number 1 represents the accesses performed by the DDM Kernel on the Meta Data Handler. The arrow tagged with number 2 represents the calls performed by the DDM Kernel to initiate and commit (or roll back)

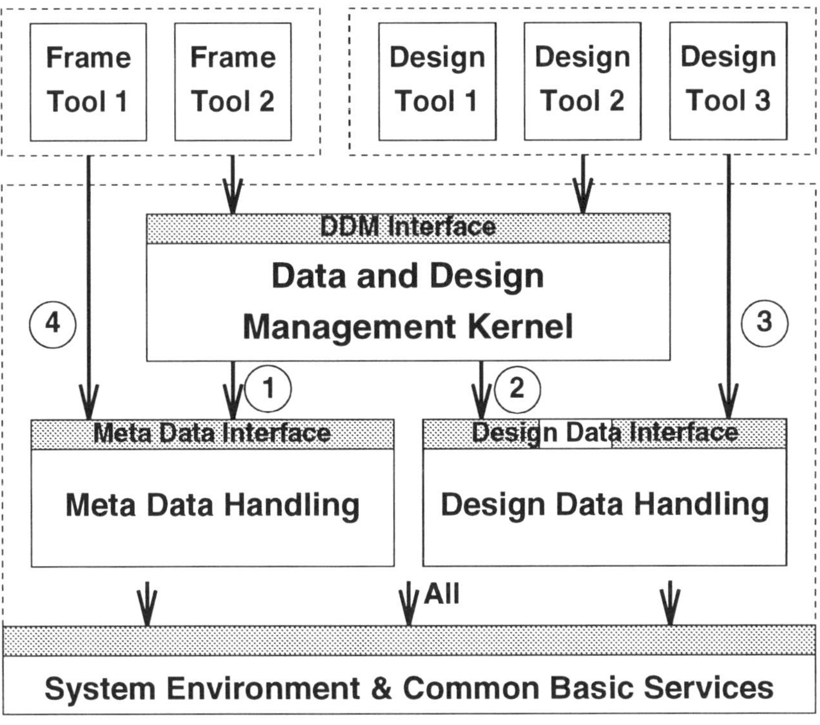

Figure 6.2 The Data Handling component is composed of separate components for Meta Data Handling and Design Data Handling.

operations on design objects. These calls are properly sequenced with the corresponding accesses performed via arrow 1, according to the procedures described above. Via arrow 1 the DDM Kernel obtains the identifications of the valid design objects that are to be accessed via arrow 2. The arrow tagged with number 3 represents the accesses by the design tools on the contents of the design objects. These are the actual storage and retrieval operations on the detailed design data. Thus, we distinguish between *design object level accesses* on the Design Data Handler (arrow 2) and actual *design data accesses* (arrow 3). This is indicated in Figure 6.2 by the two shaded areas separated by a blank part. The accesses via arrow 3 can be performed only if the design transaction on the design object in question has been initiated properly via arrow 2. The DDM Kernel performs no design data accesses. Design tools are not allowed to perform design object level accesses via arrow 3.

Compared to Figure 6.1, arrows 3 and 4 in Figure 6.2 are more specific now. Arrow 3 correctly indicates that design tools interact directly with the Data Handler to access design data only. Access of design tools to meta data is performed only via calls to the DDM Kernel. Full control of the DDM Kernel over meta data accesses by design tools is key to *consistency* of the administered state of design.

Arrow 4 correctly indicates that framework tools interact directly with the Data Handler for purposes of meta data access. If design data is involved in an operation performed by a framework tool, this is to be handled via a call to the DDM Kernel, which will also take care of the meta data - design data consistency. For example, one may think of the removal of a design object, ordered from a graphical framework browser. These are *domain neutral* operations that do not interpret detailed design data.

6.5   THE META DATA HANDLING COMPONENT

We have shown that the architecture allows separate data handling components for meta data and 'raw' design data. We will now address the Meta Data Handling component in more detail.

6.5.1   Data Types and Access Methods

The Meta Data Handling component is a facility for reliable persistent storage of meta data. It is the *framework's notebook* for maintaining the state of design, and as such it is consulted and updated frequently. In the previous chapter we derived a data schema which defines the types of information that are maintained about the design objects and their relationships. A first order approximation of the types of data that are to be handled by the Meta Data Handling component is hence given by the object types defined in the data schema.

The accesses performed on the meta data are *queries* for small amounts of data. Given a data schema, all possible queries are given by the OTO-D DML. Ultimate flexibility is hence obtained if the OTO-D DML is taken as the basis for meta data access. This does not imply that the actual DML syntax has to be supported by the meta data handling interface. Rather, the interface must allow each query that can be formulated in OTO-D DML, to be executed in some form. For example, a procedural interface may provide functions to iterate over the instances of an object type, as an equivalent of a GET request.

6.5.2   Configurability

The data schema derived in the previous chapter should not be taken as the fixed-for-all-times ideal data schema. Schema changes will result from evolving requirements. The introduction of a new framework service typically implies the introduction of new object types in the data schema. Hence, the data schema should not be hardwired in the Meta Data Handling component. Instead, we must allow the component to be *configured* with an actual data schema. A Meta Data Handling component that is completely free from data schema dependencies will facilitate framework evolution significantly. In addition it offers the possibility to allow user-defined extensions to the data schema.

6.5.3 Logical Distribution of Meta Data

An important design choice is whether to have a *single meta data repository*
for the complete environment, or to have *multiple meta data repositories*
which can be addressed individually. The former is referred to as *logical
centralization* of meta data, the latter as *logical distribution* of meta data. In
the case of logical centralization, queries are performed against all meta data
present in the environment on the basis of a *single data schema* (possibly
through some user view on this schema). An example of such an overall
data schema is the one derived in chapter 5 (see Figure 5.22). In the case of
logical distribution, a query is performed against the subset of the meta data
present in the local repository to which the query is directed. This repository
has a *local data schema* defining the structure of the information contained
in it.

In section 5.3 we introduced the notion of *project* to provide logical distribution of design data and design activities. We now consider how this
global framework aspect relates to logical distribution at the level of data
handling. Since a project provides a local context for performing design activities, which contains local design objects, most of the meta data accesses
will address the meta data that correspond to a particular project. That is,
queries will typically contain the clause:

```
WHERE ... ITS Project = '...'
```

A possible form of logical distribution is to have a *meta data repository per
project*. A query is then directed to a particular repository, and thereby implicitly to the meta data corresponding to a particular project. This matches
well with the idea of having projects to offer local environments. Such distribution of meta data over multiple smaller repositories can have a positive
impact on the *performance* and the *scalability* of the system. Moreover, it
provides *inherent parallelism* for meta data accesses on different projects,
and thereby increases the availability of the meta data. Entering a project
may imply that permissions are checked to access the meta data of that particular project. That is, logical distribution also provides an intuitive way of
controlling access to (parts of) the meta data. An important advantage of
this form of logical distribution is that it allows different projects to have
different data schemas. This implies increased flexibility, in particular if we
allow user-defined extensions to data schemas, or want to support different
variant data schemas in parallel. An obvious drawback, of course, is that
the scope of individual queries is always restricted to the meta data of a
particular project. We feel this drawback can be overcome without much

pain if we provide to the higher level components a means to interact with the meta data repositories of multiple projects in parallel.

In order to obtain logical distribution, we have to define *how* the meta data gets distributed over multiple repositories. That is, from the overall data schema defined in chapter 5 (see Figure 5.22), we have to derive the data schemas for the individual meta data repositories. This can be done systematically by studying the object types one by one. We will make some brief remarks on the data schema decomposition process.

A project meta data repository contains the meta data that belongs to the individual project. Reading from the data schema, this certainly involves the instances of the object types that (indirectly) have Project *only once* as an attribute. These object types will be part of the local project data schema, in which the Project attribute has become implicit. Examples are the object types Cell, View and DesignObject.

Special attention has to be paid to object types that may allow meta data from different projects to be related. These object types can easily be recognized since they (indirectly) have the object type Project *multiple times* as an attribute. Examples are the object types HierarchyRel, EquivalenceRel, DesignTransaction and ImportedDesignObject. In section 5.11 we defined a constraint that restricts hierarchical, equivalence and version derivation relationships to relate only design objects from the same project. The corresponding object types will, hence, also be part of the local project data schema. The object types DesignTransaction and ImportedDesignObject allow meta data from different projects to be related. Such object types have to be split into multiple object types, such that each object type again represents meta data that belongs to an individual project. For example, the object type ImportedDesignObject can be split into an object type Export, to represent the export of the design object in the exporting project, and a re-defined object type ImportedDesignObject, to represent the reference to the remote design object in the importing project. Thus, for such object types, meta data can be localized in the project environments at the cost of some redundancy. This redundancy implies that additional consistency constraints have to be safeguarded. For example, an instance of the re-defined object type ImportedDesignObject should not exist without a corresponding instance of the object type Export in the exporting project.

What remains are the object types that do *not* have Project (indirectly) as an attribute. An example is the object type Project itself. A *global meta data repository* can be allotted to this kind of information. One of the object types in the data schema for this global repository will be Project. In other words, this repository administers the projects present in the design environment. Consistency among this global administration and the actual presence of the projects in the design environment has to be safeguarded. This global repository may also contain information about the available tools, their design activities, and the configured design flows. An important design choice is whether to *reference* this tool & flow information from the project environment, or to *duplicate* information about the selected tools and design flow into the project environment. In the latter case local autonomy is enhanced, since updates performed in the global repository do not affect the selected tool & flow information in the project environment. This by itself, however, may also turn out to be a drawback, depending on the situation. We consider this choice to be a matter of taste.

We can now refine the intuitive view on the design environment presented in Figure 5.2, to reflect the logical distribution of meta data as well. See Figure 6.3.

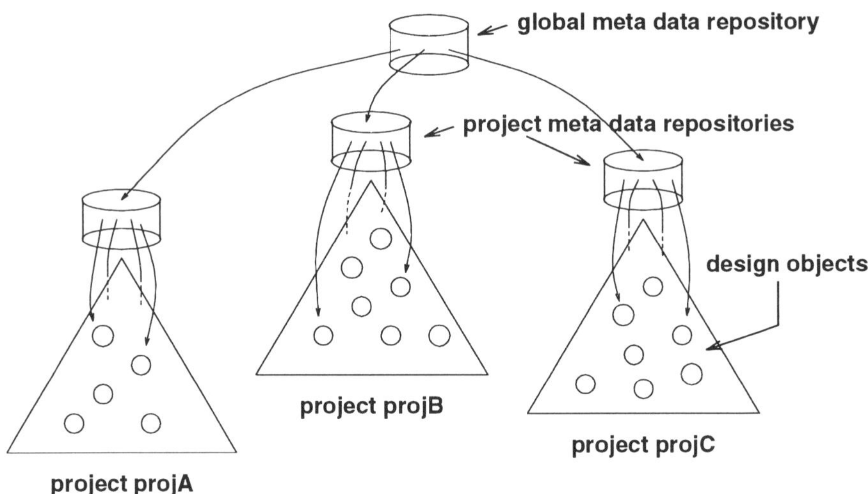

Figure 6.3 The meta data is logically distributed over local project meta data repositories and a single global meta data repository.

6.5.4 Meta Data Transactions

In section 3.2 we characterized the framework kernel as a *transaction processing system* which takes the design data from one consistent state to another consistent state. Upon the execution of the transactions initiated by tools on the framework kernel, the DDM Kernel performs accesses on the Meta Data Handler to consult and update the administered state of design. An obvious design choice is to equip the Meta Data Handler with a basic transaction facility, to allow the DDM Kernel (and other client components) to manipulate the meta data atomically. A transaction facility offers the following properties (see section 3.2):

- *Failure atomicity*: The Meta Data Handler rolls back the results of uncommitted operations when a failure occurs. Client components do not have to worry about undoing the updates that have already been performed in the course of a transaction.

- *Permanence*: Once a meta data transaction has been committed, the results will not be lost. Client components can be sure that meta data updates have been executed, and can safely return control to their clients.

- *Serializability*: Where and how should we control the concurrency of requests being issued to the framework kernel by multiple tools running in parallel? One way may be to *serialize* the requests from the tools in the DDM Kernel. From the perspective of the DDM Kernel this would free the Meta Data Handler from concurrency considerations, since the meta data transactions are executed serially. The alternative is to have the DDM Kernel handle requests from tools *in parallel*, which implies a need for allowing concurrent meta data accesses. Parallel execution of tool requests is desirable for purposes of efficiency. Besides, we allowed framework tools to interact directly with the Meta Data Handler (arrow 4 in Figure 6.2). This also raises a need for concurrency control at the level of the Meta Data Handler. We decide that the Meta Data Handler must guarantee serializability upon concurrent execution of meta data transactions.

We thus choose to equip the Meta Data Handler with a full-fledged transaction facility. The careful reader will have noticed that we already required the Meta Data Handler to offer a transaction facility when we were solving the meta data - design data consistency problem in the previous section.

What are the required characteristics of the meta data transaction facility? The framework kernel is to provide a facility for long design transactions. However, as we will show in more detail in section 6.7, this by itself does not require a facility for long transactions at the level of the Meta Data Handler. In fact, the model adopted for design transactions in section 3.4 already showed how two short operations (CheckOut & CheckIn) are used to bracket a long lasting design transaction. Since this conversational transaction model can also be applied to other cases where long transactions may occur, we feel that the Meta Data Handler may be equipped with a facility for *short transactions*. This implies that classical techniques for transaction handling, which may e.g. force transactions to wait, are appropriate. Having only short transactions helps us to satisfy the requirement for a *high availability* of the meta data.

To ensure that concurrent transactions do not interfere with each other's operation, i.e. to provide serializability, the Meta Data Handler may employ locking techniques. It may perform locking at the coarse-grain level of the *object types* defined in the data schema. That is, the complete set of instances of an object type is locked as a single object. An effective locking protocol is the following:

- At the start of a meta data transaction we lock all object types for which instances are expected to be accessed in the course of the transaction.

We emphasize that these locks are meta data locks only. They do not inhibit design data accesses for which permission has already been granted. Object types may be locked for either Read or Write. Concurrent Read - Read is allowed. Write is exclusive. The object types that are the subject of INSERT, UPDATE, or DELETE operations have to be locked for Write. Object types for which instances are accessed only by means of a GET operation or by means of an attribute traversal via the ITS construct, have to be locked for Read. Since all locks needed to perform a transaction are acquired at once, this is a two-phase locking mechanism (see again section 3.2). Two-phase read/write locking is a common technique used to guarantee serializability [Mul89].

The locking protocol presented is *efficient*, since locking is performed at the level of the object types rather than at the level of the individual instances. Another characteristic is that all object types involved have to be known at the start of the meta data transaction. In this respect we have taken considerable advantage of the fact that the DDM Kernel and the framework tools implement a pre-defined set of transaction types. As a consequence, the set of object types (possibly) involved in a meta data transaction is pre-determined. We will elaborate on this in section 6.7. A drawback of the presented protocol is that meta data transactions which acquire Write locks, tend to exclude many other transactions from operating concurrently. We note, however, that many transactions acquire only Read locks. For example, the multitude of browse operations initiated by the end-user perform only GET operations on the meta data. Another key issue is the *inherent parallelism* provided by the logical distribution of meta data. A meta data transaction is issued against an individual meta data repository. Locking is performed per meta data repository, i.e. per project. Meta data transactions issued against different meta data repositories do not exclude each other.

6.5.5 Integrity Constraints

The Meta Data Handler can make a valuable contribution to maintaining overall consistency and integrity of design information, by guaranteeing correctness of the meta data with respect to pre-defined consistency and integrity constraints. As we explained in section 4.3, the OTO-D data model allows a variety of integrity constraints to be defined. These include the *inherent constraints*, which follow directly from the application of the OTO-D DDL constructs. An example is the referential integrity along the attribute relationships. Another type of integrity constraints are the *static constraints*, of which several were defined in chapter 5 while deriving the data schema. Upon an INSERT, UPDATE, or DELETE operation, the Meta Data Handler has to check whether none of the integrity constraints is violated. When a violation is detected, the operation fails and the embracing meta data transaction must be rolled-back.

6.5.6 The Meta Data Interface Definition

We represent the various key aspects of the Meta Data Interface by means of the following interface definition. The interface definition specifies the principal functions that the Meta Data Handler makes available to the higher level components. As has been defined in the component architecture of Figure 6.2, these client components are the DDM Kernel (arrow 1) and the framework tools (arrow 4). For design tool integrators this interface is considered an *internal interface* of the CAD framework.

Two functions are used to establish and release contact with individual meta data repositories. They reflect the logical distribution of the meta data.

```
mdiOpenProject (projectID, mode): mdiProjectKey

mdiCloseProject (mdiProjectKey)
```

The function *mdiOpenProject* establishes contact with the meta data repository identified by the argument *projectID*. A reserved identifier may be used for identification of the global meta data repository. Upon activation of the identified repository, the Meta Data Handler configures itself for the corresponding data schema. The *mode* argument specifies the required access mode. The function *mdiOpenProject* checks whether meta data access with this access mode is allowed for the client issuing the request. Upon successful completion, *mdiOpenProject* hands out the access key *mdiProjectKey*, which can be used for subsequent meta data accesses on the repository.

The function *mdiCloseProject* releases contact with the meta data repository identified by the access key *mdiProjectKey*. The access key is invalidated.

Two functions are used to initiate and terminate meta data transactions on a meta data repository.

```
mdiClaimMetaData (mdiProjectKey, objectTypes,
                              lockModes): mdiTransactionKey

mdiReleaseMetaData (mdiTransactionKey, releaseMode)
```

The function *mdiClaimMetaData* initiates a meta data transaction on the repository identified by the access key *mdiProjectKey*. According to the presented locking protocol, all object types for which instances are expected to be accessed in the course of the transaction, are locked. The *objectTypes* argument specifies the set of object types. The *lockModes* argument specifies the requested lock types (Read or Write) for the locks on these object types. If not all locks can be obtained, a short wait is introduced before issuing a new attempt. The wait-retry cycle is repeated till all locks are obtained. Since *mdiClaimMetaData* holds no locks while waiting, deadlock will not occur. Upon successful completion, *mdiClaimMetaData* hands out the key *mdiTransactionKey*, which can be used to perform the actual meta data requests.

The function *mdiReleaseMetaData* terminates the meta data transaction identified by *mdiTransactionKey*. It establishes a synchronization point by performing either a commit or a rollback operation, depending on the value of the *releaseMode* argument. The locks on the object types are released. The key *mdiTransactionKey* is invalidated.

The client can obtain multiple access keys for different meta data repositories via *mdiOpenProject*. For each of these access keys, a meta data transaction can be initiated via *mdiClaimMetaData*. This interface, hence, allows parallel transactions on different meta data repositories. We note, however, that per meta data repository we allow only one meta data transaction to be in progress per client. This simplifies the locking procedure and prevents deadlock, while imposing no principal restrictions upon practical use.

Once a transaction has been initiated, actual meta data accesses can be performed with the key *mdiTransactionKey*. For this purpose we introduce the following function:

```
mdiQueryMetaData (mdiTransactionKey, queryRequest):
                                        mdiQueryResult
```

The function *mdiQueryMetaData* issues the specified *queryRequest* on the meta data repository for which the meta data transaction identified by *mdiTransactionKey* has been initiated. Upon successful completion, it returns the result of the query request via *mdiQueryResult*. *mdiQueryMetaData* checks whether the object types involved in the query request have been locked properly.

The function *mdiQueryMetaData* is presented merely to show that queries can be executed in some form in the course of a meta data transaction. It is not intended to specify *how* the interaction with the Meta Data Handler actually occurs when performing a query request. One way may be to have a procedural interface which allows clients to iterate over the instances of an object type, in order to search for instances that satisfy a search pattern representing the qualifying predicate of the query request. Selected instances are obtained one by one, to be involved in subsequent operations. An alternative is to have a more declarative form of interaction, very much like the function format specified above. This form of interaction has been used in the Nelsis CAD Framework, where the OTO-D DML was actually embedded within the C programming language. Queries in OTO-D DML syntax can be passed as strings to the Meta Data Handler. The Meta Data Handler parses the query string to an internal form which is executed efficiently. The advantage of this type of interface is its high level of abstraction. Higher level components can simply issue OTO-D DML requests, without having to wonder how these requests get resolved [vdWSBD90].

The calling pattern presented below illustrates how the Meta Data Interface functions cooperate.

```
mdiProjectKey := mdiOpenProject (projectID, mode);
   mdiTransactionKey := mdiClaimMetaData (mdiProjectKey,
                                  objectTypes, lockModes);
      mdiQueryResult := mdiQueryMetaData
                        (mdiTransactionKey, queryRequest);
   mdiReleaseMetaData (mdiTransactionKey, releaseMode);
mdiCloseProject (mdiProjectKey);
```

6.5.7 Conclusion

We conclude that the Meta Data Handler is a powerful framework component which offers a high level interface to its client framework components:

- The Meta Data Interface allows OTO-D DML requests to be issued in some form.

- It enables the meta data to be distributed logically over multiple individual repositories.

- Per meta data repository, a data schema can be configured. This configurability also allows the framework to evolve smoothly upon schema changes.

- Meta data transactions can be performed per meta data repository.

- Parallel transactions can be performed on different meta data repositories.

- It guarantees correctness of the meta data with respect to a variety of pre-defined integrity constraints.

- It handles access control.

In the next chapter we will see that physical distribution as well as recovery can be handled transparently. This makes the Meta Data Handler a powerful component for conveniently building higher level framework services.

6.6   THE DESIGN DATA HANDLING COMPONENT

We will now address the Design Data Handling component in more detail. This component provides a facility for reliable persistent storage of design data. We do *not* intend to present actual design data handling mechanisms or a detailed Design Data Interface. Instead, we will identify some global characteristics of the Design Data Handler and demonstrate how the CAD framework can provide openness to a variety of storage regimes for the 'raw' design data.

6.6.1   Data Types and Access Methods

The design data is organized as design objects contained in projects. We have defined no constraints on the amount of data or the types of data that may be contained in a design object. Design tools issue design transactions on design objects to access the design data. By means of a design transaction, a design tool may operate for a possibly long time on the design data contained in a design object.

The *type of interaction* performed by a design tool in the course of a design transaction can vary significantly from one tool to the other. Some tools prefer a very loose type of interaction with the Design Data Handler. They perform, what we call, *file level interaction*. Such tools basically want to have access to design files contained in the design object, to produce other files that are to be stored with the same or another design object.

A slightly more intimate type of interaction is found with tools that use the Design Data Interface to (sequentially) read and write the actual design data elements. This type of interaction allows the Data Handler to decouple the design tools from the actual format used for physical storage of the design data. Design tools that perform this type of interaction typically *load* the design data into their incore data structures, right after the start of the design transaction. These data structures have been optimized for the specific algorithms that the design tools are to perform. For example, analysis tools typically use very specific data structures for their particular simulation or verification algorithms. Many editors also use incore data structures for fast handling of the design data. If the tool operation completes successfully, the resulting data is *saved* (i.e. made persistent) by storing it either in a new (related) design object, or as derived data with the original design object. In case of an editor, the resulting data is stored in a new design object which may become a new version of the corresponding view.

The most intimate type of interaction is performed by design tools that manipulate the design data as it resides in the Data Handler. These tools do not build their own data structures according to the "load - manipulate - save" type of interaction presented above. They interact continuously with the Design Data Handler to navigate through the data and to perform their operations directly on the data in the Data Handler. The Data Handler is responsible for loading the design data incore and for making it persistent again. This may be done in a highly transparent way. For this type of interaction, the Design Data Handler must offer a great variety of data structuring capabilities and traversal methods. A drawback of this approach is that if multiple tools are to share the same data, they should adhere to a common data structure for performing their algorithms. If a tool does not adhere then it can, of course, still build its own data structure. However, by degrading to the "load - manipulate - save" type of interaction, it loses the advantages offered by the persistency service of the Data Handler.

6.6.2 Design Data Handling Characteristics

We have positioned the Design Data Handling component in the component architecture (see Figure 6.2) by deciding:

- to distinguish between the DDM Kernel and the Data Handler, and

- to split the Data Handler into a Meta Data Handler and a Design Data Handler.

The Design Data Handler offers functions for creating, storing, retrieving and deleting design data. In order to become more specific about the functionality offered by the Design Data Handler, we first explicitly define the discriminating principle for balancing functionality between the DDM Kernel and the Design Data Handler.

Principle 6.3 *Functions involved in design data handling are located in the DDM Kernel if they perform meta data accesses upon their execution.*

Principle 6.3 is a direct reflection of the fact that the Design Data Handler does not perform meta data accesses, as appears from Figure 6.2. All requests from design tools that imply meta data access, have to go through the DDM Kernel. These include requests for the initiation and termination of design transactions and the removal of design objects. Upon the execution of such requests, the DDM Kernel performs meta data accesses on the Meta Data Handler (arrow 1 in Figure 6.2) and design object level accesses on the Design Data Handler (arrow 2 in Figure 6.2). We conclude that principle 6.3 is a direct consequence of the split of the Data Handler; this split has detailed the Data Handler and the position of its interfaces, and thereby implicitly caused the allocation of functions to the DDM Kernel.

We now address some key data handling issues, to see how they apply to the Design Data Handling component as it has been positioned in the component architecture.

- *Logical distribution.* The design objects are logically distributed over the project environments. Each design object resides in a particular project, as was also modeled in the global data schema (Figure 5.22). A related aspect is the possibility to *configure* the Design Data Handler per project. That is, to have a *local* data schema for the design data structures and *local* settings for configurable items, when so desired.

- *Access control.* The framework controls the access to design data at the coarse-grain level of design object, for example, upon a CheckOut or RemoveDesignObject operation. For this purpose, the appropriate object types have been defined in the data schema (see for example section 5.4). The access control functions perform meta data accesses, for example to check the ownership of the design object. According to principle 6.3, the coarse-grain access control is performed by the DDM Kernel. This relieves the Design Data Handler from the principal responsibility to control access to design data.

  Design tools may only perform design data accesses (via arrow 3 in Figure 6.2) for which permission has been granted by the DDM Kernel. The Design Data Handler is welcome to provide additional safety measures to keep the tool from (accidentally or intentionally) performing unpermitted accesses. If the Design Data Handler performs access control, it must be compliant with the strategy employed by the DDM Kernel.

- *Concurrency control.* As we described in section 3.3 and 3.4, the framework controls concurrent accesses to design data at the coarse-grain level of design object. Upon initiation of an operation on a design object, it decides whether this operation is allowed with respect to operations that are in progress. In chapter 5 we modeled the object type RunningDesTrans, to allow the framework to register the running design transactions in the meta data. According to principle 6.3, the coarse-grain concurrency control is performed by the DDM Kernel. This relieves the Design Data Handler from the principal responsibility to provide concurrency control.

A requirement on the Design Data Handler is that it efficiently permits concurrent execution of operations for which permission has been granted by the DDM Kernel. For example, a Design Data Handler which maps all design data present in a project onto a single file, inherently constrains concurrent access (the file is a single resource with exclusive Write access). If a single file is to be used for physical storage, some mechanism has to be built on top of it to correctly handle the concurrent accesses on the different design objects in the file. In contrast, a Design Data Handler which maps each design object onto a separate file, inherently allows concurrent access on different design objects, taking advantage of facilities provided by the underlying operating system.

- *Consistency.* The framework kernel does not interpret the contents of design objects. Hence, it is the design tool's responsibility to deliver a *self-consistent* design object. In order to provide meta data - design data consistency, we required the Design Data Handler to offer the property of *permanence* (see also sub-section 6.4.2).

In sub-section 6.4.2, we came to a distinction between *design object level accesses* on the Design Data Handler (arrow 2 in Figure 6.2) and actual *design data accesses* on the contents of the design objects (arrow 3 in Figure 6.2). We will study these two types of accesses separately. We will start with the definition of an interface for the design object level accesses. We recall that in section 3.3 the need for a "domain neutral interface for design object manipulation by the framework" was already identified.

6.6.3   Design Object Level Interface

We present the principal functions of the *design object level interface* of the Design Data Handling component. These functions are used solely by the DDM Kernel (arrow 2 in Figure 6.2). For design tool integrators this interface is considered an *internal interface* of the CAD framework.

Two functions are used to initiate and terminate design data accesses on a project environment. They reflect the logical distribution of design data across project environments.

```
ddiOpenProject (projectID, mode): ddiProjectKey

ddiCloseProject (ddiProjectKey)
```

The function *ddiOpenProject* initiates design data accesses on the project identified by the argument *projectID*. The *mode* argument specifies the required access mode. Upon successful completion, *ddiOpenProject* hands out the access key *ddiProjectKey*, which can be used for subsequent design data accesses on the project. The function *ddiCloseProject* terminates design data accesses on the project identified by the access key *ddiProjectKey*. The access key is invalidated.

In sub-section 6.4.2, we discussed the interaction of the DDM Kernel with the Meta Data Handler (arrow 1 in Figure 6.2) and the Design Data Handler (arrow 2 in Figure 6.2) in the context of meta data - design data consistency. We defined procedures for the creation, update, and removal of design objects. These procedures describe the accesses on both Data Handlers and how they have to be sequenced. Starting from the description of the accesses, we define the following functions of the design object level interface.

```
ddiCreateDesignObject (ddiProjectKey):

                                      ddiDesignObjectKey
```

The function *ddiCreateDesignObject* creates an 'empty' physical design object in the project identified by *ddiProjectKey*. The access key *ddiDesignObjectKey* is returned to permit further accesses on the new design object. This function is called upon a CheckOut for the creation of a new design object.

```
ddiOpenDesignObject (ddiProjectKey, designObjectID,
                                accMode): ddiDesignObjectKey
```

The function *ddiOpenDesignObject* opens the physical design object identified by the argument *designObjectID* in the project identified by *ddiProjectKey*. The *accMode* argument specifies the type of access that is to be performed on the design object. Compare this argument to the AccMode attribute in the definition of the object type DesignTransaction (section 5.5). If updates are to be performed on the design object, a *copy* is generated. Upon successful completion, the access key *ddiDesignObjectKey* is returned, and the Design Data Handler is prepared to accept further accesses on the (copied) design object. This function is called upon a CheckOut for the update of an existing design object.

The copy operations performed by *ddiOpenDesignObject* are potential performance killers. We like to note here that in practice the physical copying of design data can be reduced significantly. First, ReadOnly design transactions require no copying. For update operations, smart techniques may be employed. For example, we may only copy *references* to the design data contained in the original design object. Read operations get directed to the original object, while write operations are directed to the new object. It is possible to combine this in a smart way with the versioning procedures, as has been demonstrated in the Nelsis CAD Framework. We will not discuss this in detail.

```
ddiCloseDesignObject (ddiDesignObjectKey, complMode)
```

The function *ddiCloseDesignObject* closes the physical design object identified by the argument *ddiDesignObjectKey*, obtained from either the function *ddiCreateDesignObject* or *ddiOpenDesignObject*. The behavior of *ddiClose-DesignObject* is directed by the argument *complMode*. Compare this argument to the ComplMode attribute in the definition of the object type Design-Transaction (section 5.5). This function is called upon a Checkln, i.e. upon termination of a design transaction. If an empty design object or a copy design object was generated at the start of the design transaction, this design object is removed if *complMode* signals failure, and this new design object is committed if *complMode* signals success. Since we required the Design Data Handler to offer the property of *permanence*, the new design data will not be lost, and it can safely be administered in the meta data. The access key *ddiDesignObjectKey* is invalidated.

```
ddiRemoveDesignObject (ddiProjectKey, designObjectID)
```

The function *ddiRemoveDesignObject* removes the physical design object identified by the argument *designObjectID* from the project identified by *ddiProjectKey*. This function is called upon a RemoveDesignObject operation, after the removal of the design object has been administered in the meta data. It is also called upon a Checkln of a successful update of an existing design object, if the original design object is no longer to be retained (i.e. it is overwritten). It is then called after the successful completion of the design transaction has been administered in the meta data.

6.6.4   Openness to Alternative Design Data Handlers

One approach to design data handling is to require the CAD framework to offer a comprehensive facility which supports the great variety of data types and access methods needed by the possible applications. All design tools that are to be integrated, are obliged to interface to this facility for the handling of their design data. This is the approach followed by the Jessi-Common-Framework [JCF91b]. There are, however, several reasons to diverge from this approach.

- We think that a single data handler, which efficiently satisfies the disparate needs of *all* possible applications, is hard to build. It will be a complex facility that has to compromise between the many (conflicting) requirements posed on it. *Run-time efficiency* will be hard to achieve.

- Many existing tools use 'their own' facilities for design data handling. Requiring these tools to interface to the unique framework facility for the handling of their design data, may severely complicate incorporation of these tools into the framework based design environment. In section 3.3 we stated that the framework must allow incorporation of a design tool through integration at the design environment level only. This is an important aspect of *openness*.

- Putting all cards on a single design data handler, excludes the possibility to incorporate state-of-the-art data handling technology developed elsewhere. In the commercial arena, object-oriented database management systems are hitting the market. Some examples are the Objectivity/DB product from Objectivity, Inc [Obj91] and ObjectStore from Object Design, Inc. Looking one step further ahead, object management systems are expected to be commodities several years from now, to be delivered as 'standard' components with the operating system. In our component architecture this can be seen as a migration of data handling functionality into the System Environment and Common Basic Services component.

A flexible approach to design data handling is to allow CAD tool integrators to incorporate alternative Design Data Handlers. Such flexibility can be provided by defining an appropriate interface for the interaction between a data handler and the other framework components. An alternative data handler can then be incorporated by equipping it with this interface, provided it satisfies additional requirements for the correct cooperation with the other parts of the framework.

In sub-section 6.4.2, we concluded that the Design Data Handling component has a *design object level interface*, to be used by the DDM Kernel (arrow 2 in Figure 6.2), and a *design data access interface* to be used directly by design tools (arrow 3 in Figure 6.2). The only direct interaction between a framework component and the Design Data Handler, thus, occurs via the design object level interface, which we defined above. This interface may become the standard interface for allowing alternative Design Data Handlers to be incorporated. This is represented by the component architecture in Figure 6.4.

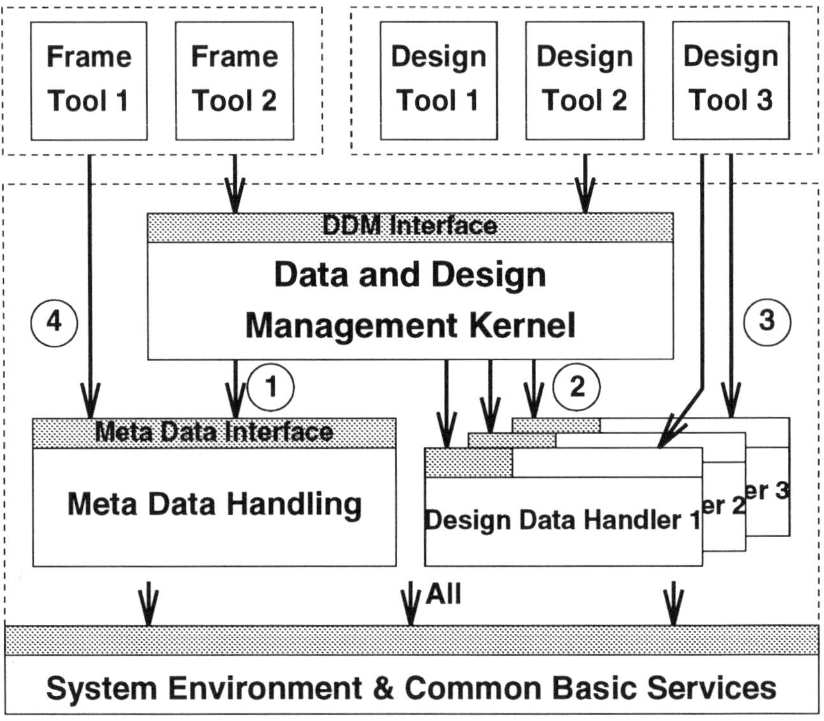

Figure 6.4 The component architecture allows alternative Design Data Handling components to be incorporated.

The different Design Data Handlers in Figure 6.4 have all been equipped with the standard design object level interface. Via this standard interface, the DDM Kernel can perform its design object level accesses. These accesses follow the procedures defined in sub-section 6.4.2, thereby guaranteeing meta data - design data consistency.

Design tools must interact with the DDM Kernel to initiate and terminate design transactions. In response to a successful CheckOut request, the DDM Kernel returns an access key for the selected design object (as we will see in more detail in section 6.8). With this key, the design tool can contact the appropriate Data Handler, in order to perform subsequent design data accesses.

The obvious primary requirement for a data handler that is to be incorporated, is that it can be equipped with the design object level interface. It must allow design data to be identified, created, updated and removed on a *per design object* basis. In addition, we list the following requirements:

- As we discussed, the Design Data Handler must offer the property of permanence.

- The access control and concurrency control strategy employed by the Design Data Handler must be compliant with the strategy employed by the DDM Kernel. As we discussed, the DDM Kernel is the principal authority in access and concurrency control matters. The Design Data Handler must permit data accesses which have been granted by the DDM Kernel. As an additional safety measure, it may obstruct data accesses which have not been granted by the DDM Kernel.

- The data handler has to provide an interface function that derives from the DDM access key a handle suited for performing data access on the corresponding physical design object.

The openness to alternative Design Data Handlers relieves the CAD framework from the burden to offer a comprehensive facility that satisfies everyone's needs. It significantly helps the CAD framework to be a *light-weight facility*, with a focus on data and design management functionality. One or more Design Data Handlers may, of course, be provided by the CAD framework itself. By nature, these are generic facilities, as discussed in section 3.3.

Even though the design tools may employ different Design Data Handlers, we can still term the resulting system *integrated*. Integration is achieved at the level of coarse-grain design object manipulations. The framework kernel keeps track of the state of design by collecting meta data about the design objects. This meta data is administered in a single Meta Data Handling facility. From the perspective of the design engineer, who sees the meta data, all design objects reside in a single environment. He observes no redundancies or inconsistencies.

Tool interoperability also requires integration at the design representation level (section 3.3). The CAD framework does not enforce such integration. Tool interoperability primarily is a tool responsibility and should be driven by standardization. Instead, the framework pursues openness at the design data access side. Such openness also is to facilitate the incorporation of design representation formats once they get standardized. The support for alternative Design Data Handlers contributes to this openness since it causes the framework *not* to have a single interface for design data access. CFI is actively addressing standardization for the sake of tool interoperability.

Support for specific design representation formats may be provided in several different ways. One way may be to hardwire the formats in a dedicated data handler, and incorporate this handler in the system. An alternative is to have a generic component that can be configured by means of a data schema defining the structure of the data to be handled. One may also think of automatic generation of dedicated procedural interfaces from such data schemas.

An obvious way of providing domain specific extensions to the CAD framework, is by implementing a layer on top of it. This layer may also hide the DDM Interface from the tools. By means of such a layer and the use of the appropriate data handler, the CAD framework can be transformed into the integration platform for a set of tools that already employs a 'standard' interface. In the context of the Nelsis CAD Framework, this approach was used to implement the CFI DR-PI (Design Representation Programming Interface: Electrical Connectivity) [CFI92b]. This exercise is described in [Sim93].

6.6.5 Conclusion

We have identified some global characteristics of the Design Data Handler and demonstrated the openness of the CAD framework to a variety of storage regimes for the 'raw' design data. The architecture provides a uniform mechanism for meta data - design data consistency. For each application the optimal data handling component can be incorporated, tuned to the characteristics of the data to be handled. Openness is increased for design tools that employ a data handler not known to the framework. Moreover, the framework can easily evolve to incorporate state-of-the-art data handling technology developed elsewhere.

6.7 THE DATA AND DESIGN MANAGEMENT KERNEL

In this section we will further detail the Data and Design Management Kernel (DDM Kernel). We do *not* intend to present the detailed functionality of the DDM Kernel. As we concluded in section 3.5, this detailed functionality is often a matter of taste. Instead, we will identify the major characteristics of the DDM Kernel, and clarify how DDM Kernel functions relate to the information architecture derived in chapter 5.

6.7.1 Global Charter

Tools interact with the DDM Kernel to obtain access to design objects and their relationships. The principal task of the DDM Kernel is to *keep track* of the *state of design* as this interaction takes place, and to use the knowledge of the state of design to enforce constraints on the design process. This makes the DDM Kernel the *central authority* of the system. The state of design is administered in the Meta Data Handling component, structured according to a data schema.

6.7.2 Single Logical Component

We consider how to decompose the DDM Kernel into multiple logical components. In the CFI framework architecture reference (see again Figure 2.4), we see separate components for data management and methodology management, with a dependency of the methodology management component on the data management component. Tools interact separately with the data management component and the methodology management component.

We deviate from the CFI framework architecture reference, in that we want to have a *single interface* of the DDM Kernel towards the design tools. Design tools want access to design data, to perform their specific operations. They need a simple, well-structured interface, geared at design data access. This interface must hide many of the framework specifics. This will be detailed in section 6.8.

In addition, we feel that the clear separation between data management and methodology management is an over-simplification. We share the view of van den Hamer and Treffers that the topics of design data management and design process management are very closely interrelated [vdHT90, vdH91].

A single set of concepts must handle both areas in a consistent way. This can be grasped by realizing that design process management is involved with *data derivation*, for example, using a data flow paradigm (section 5.12). The concepts employed for design process management must, therefore, connect well with the data organization concepts employed for purposes of design data management. As van den Hamer says in [vdHT90]: "Failure to do so leads to a considerable increase in overall system complexity with obvious consequences for implementation, maintenance and usage of the system."

Our own experience with the Nelsis CAD Framework strengthens this view. In Nelsis, the concepts employed for design flow management directly relate to the concepts employed at the data management side for the organization of the design data. This approach also enhances the coherence of the system to the end-user. For example, when a design engineer sees in his flow browser that the edit activity has produced object B from object A, this corresponds to the version derivation between the objects A and B displayed in his version browser. In terms of component structure, our experience has been that design flow management was incorporated in a very modular way [BtBvdW92]. However, it did not turn out to be a simple "one component on top of the other" structure, as suggested by the CFI framework architecture reference.

We conclude that we do not further decompose the DDM Kernel into smaller logical components. This would imply the presentation of many low level design choices. The amount of detail involved in this is beyond the scope of this book. Instead, we will discuss the relationship between the DDM Kernel functions and the information architecture derived in chapter 5. This will provide additional insight in the internals of the DDM Kernel.

6.7.3   DDM Kernel Operations as Meta Data Transactions

Tools can issue requests to the DDM Kernel by calling DDM functions exported via the DDM Interface. While handling a request, the DDM Kernel interacts with the Meta Data Handling component, to consult and update the administered state of design. As we explained in section 6.5, the meta data is accessed by issuing OTO-D DML queries in some form. A data schema defines the information structure of the meta data. The queries that may be issued to the Meta Data Handler depend upon the actual data schema in use. We portray the DDM Kernel as a box where requests from tools go in at the top, and actual queries come out at the bottom. See Figure 6.5.

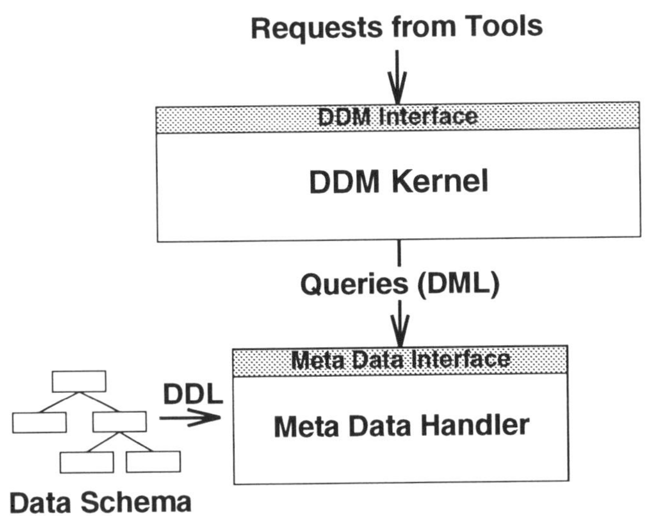

Figure 6.5 In response to requests from tools, the DDM Kernel issues queries to the Meta Data Handler.

Note that the component architecture reflects the approach we have taken to the definition of the CAD framework architecture. We started to define the *structure of the information* about the design and the design activities, without specifying the *specific functions* that have to operate on this information. Operations on the information could be expressed as DML queries. In the component architecture, the information is handled by a generic Meta Data Handling component. The specific functions are provided on top of this, and operate on the information via DML queries.

In order to learn more about the DDM Kernel, we study the relationship between the provided DDM Interface functions (DDMI functions) and the actual queries. In chapter 5 we saw that each framework service relates to one or more object types in the data schema. Multiple DDMI functions may be involved in providing a particular framework service. For example, to provide proper access control, permissions have to be checked upon a Check-Out request as well as upon a RemoveDesignObject request. For purposes of design flow management, CheckOut has to check the flow status of the design object at the start of the design transaction, and CheckIn has to update the flow status at the end. An individual DDMI function typically performs multiple queries, corresponding to different framework services.

Multiple tools may be running in parallel, issuing requests to the DDM Kernel. Upon a request, the DDM Kernel must atomically take the administered state of design from one consistent state to another consistent state. In section 6.5 we already stated that parallel execution of requests is desirable for purposes of efficiency, and we equipped the Meta Data Handler with a full-fledged transaction facility. As a result, the DDM Kernel can implement the DDMI functions as *meta data transactions* on the Meta Data Handler. A DDMI function has the following structure:

```
ddmiFunctionX ( ... )
{
    ...

    ddi...DesignObject ( ... );

    ...

    mdiTransactionKey = mdiClaimMetaData (mdiProjectKey,
                                        types, modes);

    ...

    mdiQueryResult = mdiQueryMetaData (mdiTransactionKey,
                                        query1);

    /* process result of query1. */

    ...

    ...

    mdiQueryResult = mdiQueryMetaData (mdiTransactionKey,
                                        queryN);

    /* process result of queryN. */

    ...

    mdiReleaseMetaData (mdiTransactionKey, rMode);

    ...

    ddi...DesignObject ( ... );

    ...
}
```

Due to the high level interface provided by the Meta Data Handler, DDMI functions can be implemented conveniently. We make the following remarks:

- For some DDMI functions, appropriate calls to the Design Data Handler have to be performed before and/or after the meta data transaction (the ddi...DesignObject(...) function calls in the code segment above). The procedures for meta data - design data consistency were discussed in sub-section 6.4.2 and the Design Data Handler Interface was presented in section 6.6.

- Since the DDM Interface offers a fixed set of DDMI functions to the design tools, the DDM Kernel implements a fixed set of meta data transactions. This allows locking to be performed according to the simple locking protocol presented in section 6.5.

- Introduction of a new framework service does not necessarily imply the introduction of new DDMI functions. If the available DDMI functions can correctly provide the framework service, there is no need to bother the tools with an extension of the DDM Interface definition.

- The following steps have to be taken to incorporate a new framework service:

  1. Define or refine the appropriate object types in the data schema, i.e. adjust the information architecture.
  2. Define functions that have to be performed by the framework service, expressed as DML queries on the related object types.
  3. Incorporate the queries in (existing or new) DDMI functions.

In the Nelsis CAD Framework, the presented structure of the DDMI functions is directly reflected by the structure of the source code. Most Nelsis DDMI functions are nothing more than a few embedded OTO-D DML queries with proper error handling, surrounded by a claim - release pair to ensure atomicity. This also illustrates that for many framework services the information structure dominates the algorithmic aspects of their operation, as stated in section 3.5.

6.7.4   Data Schema Dependence

A particular framework release implements a particular set of framework services on the basis of a particular data schema. In section 6.5, we presented a configurable Meta Data Handler. Upon framework evolution, this component does not have to be modified; it can simply be fed a new data schema. For a particular framework release, the data schema has to match the services implemented in the DDM Kernel. Object types may not be deleted, since the DDM Kernel performs queries on them. The data schema may, however, be extended with new object types.

6.7.5   Long Design Transactions

Using the *short* transaction facility of the Meta Data Handler, individual DDMI functions can be implemented conveniently as atomic operations. The DDM Kernel must also provide a facility for *long design transactions*. In section 3.4, we presented a conversational transaction model for design transactions. A design transaction is initiated by a CheckOut request and terminated by a CheckIn request, with a locked status in between.

The CheckOut and CheckIn operations are made available as DDMI functions to the design tools. Upon a CheckOut request, the meta data transaction queries for conflicting running design transactions, to decide whether concurrent access is allowed. If access is allowed, the new in-progress design transaction is registered in the meta data by another query within the same meta data transaction. This *locks* the design object. The underlying principle is that we serialize all requests for access to particular objects (design objects) via an authority (DDM Kernel) which maintains a lock administration (meta data) about these objects. In our data schema (Figure 5.22), we have modeled the object type RunningDesTrans for this purpose. Upon conflict, the request fails and control is returned to the design tool.

After a successful CheckOut, the subsequent accesses on the Design Data Handler (arrow 2 and 3 in Figure 6.2) operate under the protection of the lock registered in the meta data (via arrow 1). The meta data transaction performed upon the corresponding CheckIn request, removes the lock. Since the running design transactions are registered by means of non-volatile locks, the framework always knows which ones were in progress upon a tool crash. This enables the system to roll-back these design transactions.

6.8   THE DATA AND DESIGN MANAGEMENT INTERFACE

6.8.1   Introduction

In this section we define the DDM Interface, which allows tools to issue requests to the DDM Kernel. For this purpose, it makes a set of functions available: the DDMI functions. The objective of the DDM Interface is two-fold. It must provide a procedure to access design objects, and it must allow design tools to operate in a controlled way on the meta data. This is a direct consequence of the component architecture. All design object level accesses on the Data Handler (arrow 2 in Figure 6.2) have to go through the DDM Kernel, as well as the meta data accesses for design tools.

A key requirement for the DDM Interface is that it *decouples* the development and evolution of the framework on the one hand and the design tools on the other hand.

- The DDM Interface must not be tailored to specific tool features or design methodologies. For the sake of *openness*, it must be universal, to allow any type of tool to interact with the DDM Kernel.

- The DDM Interface must not be tailored to specific functionality of the DDM Kernel. For example, it must allow interfacing to framework releases with or without concurrency control, access control, etc. This will allow tools to be "plugged in" in the same way in different framework releases.

We must realize that the primary objective of a design tool upon interaction with the DDM Kernel is *not* to have data and design management functions performed. The primary objective of a tool is to obtain access to data, in order to perform its specific operations on it. Thus, instead of plainly showcasing all our fantastic data and design management functions as DDMI functions via the DDM Interface, we must hide them underneath a simple interface geared at design data access.

6.8.2   The DDMI Transaction Schema

To arrive at a simple and stable procedure for design data access, we start from the global concepts that we have used for the organization of the design data. Design data is contained in *design objects*, which reside in *projects*. A *transaction schema* defines a procedure that tools must follow to obtain access to data. Based on the hierarchical organization of the design data across design objects and projects, we adopt a layered transaction schema, which defines that tools may obtain access to design data by issuing nested transactions[1]. This layered transaction schema provides a stable backbone for the DDM Interface. Summarizing, we employ the following principle [vdMvLvdW+87]:

Principle 6.4 *We decouple the development and evolution of the framework and the design tools by adopting a common* layered transaction schema *and using this as the stable backbone for the DDM Interface.*

We adopt the following transaction schema. The effect of a tool run on a design environment is called a *tool execution*. It is a (possibly interleaved) sequence of *project transactions* bracketed by an *initialize* and a *terminate*. Similarly, a project transaction is a (possibly interleaved) sequence of *design transactions* bracketed by an *open project* and a *close project*. A design transaction is a sequence of *design data operations* bracketed by a *CheckOut* and a *CheckIn*.

We present the transaction schema graphically in Figure 6.6, using a variant of the graphical notation defined by Jackson in [Jac83]. Jackson uses this notation for so called structure diagrams, which represent the ordering of actions in time. In Figure 6.6, the rounded boxes at one level represent a sequence of actions, executed from the left to the right. Child actions specify a refinement of the parent action. A star represents iteration.

The transaction schema defines the backbone of the DDM Interface. Basically, there is one DDMI function for each leaf of the tree in Figure 6.6 (except for the design data operations, which we will not further discuss here). Access to either the design environment, a project or a design object can be obtained by executing the corresponding opening-bracket function, as represented by the leaves at the left-hand side of the tree in Figure 6.6. A transaction is terminated by executing the corresponding closing-bracket function at the right-hand side. In between, lower-level transactions can be performed.

---

[1] We use the term *transaction* in this section to indicate that operations at a particular level are bracketed by a *begin*-marker and an *end*-marker. However, this is not intended to imply all properties of the transaction concept as defined in section 3.2.

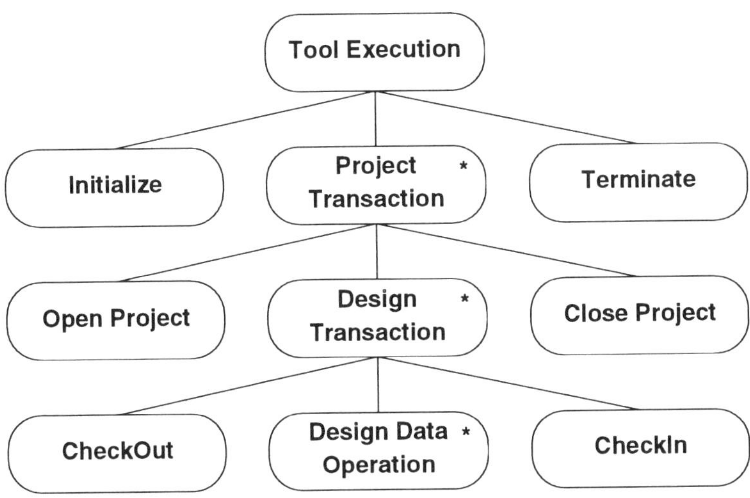

Figure 6.6   Layered transaction schema of the DDM Interface.

The functions cooperate with each other in such a way that access proceeds in accordance with the transaction schema. For this purpose we employ a *key-mechanism*. There are three types of keys, one for each layer:

$$
\begin{array}{lcl}
\text{DDMI\_ENVIRONMENT} & \Longleftarrow & \text{environment transaction key,} \\
\text{DDMI\_PROJECT} & \Longleftarrow & \text{project transaction key,} \\
\text{DDMI\_DESTRANS} & \Longleftarrow & \text{design transaction key.}
\end{array}
$$

The key returned by an opening-bracket function at some layer is part of the argument list of the functions at the next lower level and of the closing-bracket function. This allows the interleaving of more than one sequence of calls of lower-level functions. The closing-bracket function invalidates the key.

Typically, a key contains information about the object for which access was obtained, for use by the lower-level functions. Each key also contains a pointer to the next higher level key that was passed as an argument to the function returning the lower level key, so that the complete context is known at the lowest levels. Also, all keys with the same parent key are linked together in a list that is attached to this parent key. This facilitates error handling and automatic clean-up actions. When a key is invalidated by the corresponding closing-bracket function, it is removed from the list.

The transaction schema of Figure 6.6 provides a well-structured backbone for the DDM Interface. The DDMI functions derived from it, *localize* the interaction between the tools and the DDM Kernel for the access to design data. DDM services can be associated with these functions. For example, if some form of access control is to be performed at the project level, this can be done transparently underneath the OpenProject function. Logical distribution can be handled at the project level. Concurrency control, version management and design flow management can be handled at the design transaction level. Since the key-mechanism allows information to be communicated transparently across the different levels, visibility of particular DDM features is confined to a small number of places.

6.8.3 The DDMI Functions

We start with the presentation of the DDMI functions that follow from the transaction schema. Two functions take care of global initialization and termination. They establish and release contact between the tool and the design environment.

```
ddmiInitialize (toolName, options): ddmiEnvironmentKey

ddmiTerminate (ddmiEnvironmentKey, termMode)
```

Function *ddmiInitialize* is the opening-bracket function of a tool execution and returns a DDMI_ENVIRONMENT key. This key contains information about the environment in which the tool is executed (for example, hostname, user-id, working directory, etc.). The tool identifies itself by means of the argument *toolName* and informs the DDM Kernel about the *options* used to run the tool. This function may perform, for example, a license check.

Function *ddmiTerminate* is the closing-bracket function of a tool execution. It takes care of the necessary clean up operations. The behavior of ddmiTerminate is directed by the argument *termMode*.

In between ddmiInitialize and ddmiTerminate, project transactions may be executed. At this level the framework may handle such aspects as logical distribution, physical distribution and access control.

```
ddmiOpenProject (ddmiEnvironmentKey, projectID,
                                opMode): ddmiProjectKey

ddmiCloseProject (ddmiProjectKey, cpMode)
```

Function *ddmiOpenProject* initiates a project transaction. It returns a DDMI-
PROJECT key. This key contains information about the particular project,
identified by the argument *projectID*, and the access mode, represented by
the argument *opMode*. For the requesting tool, ddmiOpenProject contacts
the Meta Data Handler, via mdiOpenProject (section 6.5), and the Design
Data Handler, via ddiOpenProject (section 6.6). A project server may be
activated and access rights may be checked. The project key can be passed
as an argument to the functions at the design transaction layer.

Function *ddmiCloseProject* terminates the project transaction identified by
the argument *ddmiProjectKey*. The behavior of ddmiCloseProject is directed
by the argument *cpMode*.

In between ddmiOpenProject and ddmiCloseProject, design transactions may
be executed. The functions at the design transaction level may take care of
such aspects as versioning, view types, design flow management, etc.

```
ddmiCheckOut (ddmiProjectKey, designObjectID, accMode):
                              ddmiDesignTransactionKey

ddmiCheckIn (ddmiDesignTransactionKey, complMode)
```

Function *ddmiCheckOut* is the opening-bracket function of a design transaction. Its arguments are a DDMI_PROJECT key, identifying a project for
which access rights have been obtained via ddmiOpenProject, and an identification of a design object, denoted by *designObjectID*. The *accMode* argument
specifies the type of access that is to take place. Compare this argument to
the AccMode attribute in the definition of the object type DesignTransaction (section 5.5). Function ddmiCheckOut calls the Design Data Handler to
perform either a ddiCreateDesignObject or a ddiOpenDesignObject (section
6.6).

Function *ddmiCheckIn* terminates the design transaction identified by the
argument *ddmiDesignTransactionKey*. The behavior of ddmiCheckIn is controlled by the argument *complMode*. Compare this argument to the ComplMode attribute in the definition of the object type DesignTransaction (section 5.5). Function ddmiCheckIn calls the Design Data Handler to perform a
ddiCloseDesignObject and possibly a ddiRemoveDesignObject (section 6.6).

In addition to the DDMI functions that follow directly from the transaction
schema, the DDM Interface offers various other functions. Each of these
functions fits in the access procedure implied by the transaction schema. We
present some important functions.

```
ddmiSelectDesignObject (ddmiProjectKey, name, viewType,
                      vNumber, vStatus): designObjectID
```

Function *ddmiSelectDesignObject* may be used to obtain the identification of a design object in the project identified by the argument *ddmiProjectKey*. This object identification may be passed to other functions, such as ddmiCheckOut. This function provides the service of *name resolution*. From a name and a view type, and (optional) values for the version number and/or version status, it identifies an individual design object.

```
ddmiRemoveDesignObject (ddmiProjectKey, designObjectID)
```

Function *ddmiRemoveDesignObject* removes the design object identified by the argument *designObjectID* from the project identified by the argument *ddmiProjectKey*. The meta data transaction performed by this function checks whether removal is allowed, in particular with respect to the relationships in which the design object is involved. Note that the inherent constraints of the OTO-D data model prevent the deletion of an instance of the object type DesignObject if the (hierarchical or equivalence) relationships on it have not been removed first. Function ddmiRemoveDesignObject calls the Design Data Handler to perform a ddiRemoveDesignObject (section 6.6).

```
ddmiInstallDesignObject (ddmiProjectKey,
              oldDesignObjectID, newDesignObjectID, ...)
```

Function *ddmiInstallDesignObject* performs the *install operation* discussed in section 5.10. In the project identified by the argument *ddmiProjectKey*, the design object identified by the argument *newDesignObjectID* is substituted in the design hierarchy for the design object identified by the argument *oldDesignObjectID*.

The functions presented above, allow design tools to operate on design objects. In addition, functions must be provided which allow design tools to operate in a controlled way on the meta data. The object types of interest to design tools can be identified from the data schema. For these object types, the appropriate operations can be defined and made available as DDMI functions to the design tools. We present some examples.

```
ddmiPutEquivalence (ddmiProjectKey,
        sourceDesignObjectID, targetDesignObjectID, class)
```

```
ddmiGetEquivalence (ddmiProjectKey, designObjectID,
                      class, viewType, ...): EqRelList
```

Function *ddmiPutEquivalence* allows a design tool to establish an equivalence relationship between the design objects identified by the arguments *sourceDesignObjectID* and *targetDesignObjectID*. The equivalence class is specified via the argument *class*.

Function *ddmiGetEquivalence* retrieves equivalence relationships for the design object identified by the argument *designObjectID* in the project identified by the argument *ddmiProjectKey*. Selection of the equivalence relationships can be controlled via several arguments, such as *class* and *viewType*. Information on the selected equivalence relationships is returned in the form of a list of structures.

```
ddmiPutHierarchy (ddmiDesignTransactionKey,
                   designObjectID, instName, constructor)
ddmiGetHierarchy (ddmiDesignTransactionKey):
                                              HierRelList
```

Function *ddmiPutHierarchy* allows a design tool to establish a hierarchical relationship. The parent design object is the new object that is being produced by the design transaction identified by the argument *ddmiDesignTransaction-Key*. The child design object is identified by the argument *designObjectID*. The arguments *instName* and *constructor* specify the values for the other attributes of the object type HierarchyRel.

Function *ddmiGetHierarchy* retrieves the hierarchical relationships that have the design object on which the design transaction identified by the argument *ddmiDesignTransactionKey* has been initiated as the parent design object. Information on the hierarchical relationships is returned in the form of a list of structures, where each structure contains the identification of the child design object, the related instance name and the constructor.

We see that access on hierarchical relationships can be performed only by initiating a design transaction on the parent design object. New hierarchical relationships can be established only by creating a new design object (ddmiInstallDesignObject allows existing hierarchical relationships to be redirected). Retrieval of hierarchical relationships can be performed only for objects that have been locked. As explained in section 5.10, such a lock inherently 'freezes' the design hierarchy in the downward direction. These measures prevent inconsistencies due to concurrent operation of design tools on design hierarchies.

6.8.4 The Nelsis DMI

The approach to the definition of the DDM Interface presented here, has also been followed by the Nelsis CAD Framework. The Nelsis DDM Interface is named *Data Management Interface* (DMI), for historical reasons [vdMvLvdW$^+$87]. The three top levels of the DMI transaction schema are identical to the transaction schema of Figure 6.6. The DMI transaction schema has one additional level: the *stream transactions*.

Nelsis employs the domain neutral concept of *stream*, to have some control over the accesses that design tools perform on the contents of design objects. The contents of a design object is to be organized as one or more streams, and access on streams is to be initiated and terminated via the appropriate DMI functions. A stream may correspond to a file. In our discussion on CAD framework architectures, we have (on purpose) not addressed this level of detail. However, proper design choices at this level are crucial upon realization of actual data and design management services. For example, Nelsis exploits the stream access information in the area of design flow management. The design activities that a tool performs upon a tool run, are recognized on the basis of the stream accesses performed in the course of the design transactions. For this purpose Nelsis employs an extension of the data schema to allow the configuration of stream types and their relationship to data types:

```
TYPE StreamType = StrName, DataType
```

That is, a number of streams can be handled as a single data item of a particular data type in the context of design flow management. By monitoring the stream accesses, the framework learns which data types are accessed in the course of a design transaction.

The Nelsis DMI was introduced in 1986, and was employed by over 50 design tools to interface to the Nelsis CAD Framework. Since 1986, the framework implementation has evolved from a simple single-user single-host data management system to the multi-user distributed system that it is today, offering advanced data and design management services. In the course of this evolution, the Nelsis DMI had to undergo only a few minor changes. Upon a new framework release, design tools were 're-integrated' simply by linking them to a new release of the DMI function library.

6.8.5   Integration and Encapsulation

An aspect of tool integration is the actual *technique* used to perform the integration. In the field of electronic design automation one typically distinguishes between *tight integration*, or simply integration, and *encapsulation* [CFI90b]. Upon tight integration, the source code of the design tool is modified, to include code that handles the interaction with the CAD framework. Upon encapsulation, the source code of the design tool is not modified. Instead, a *wrapper* of additional code is written, to loosely interface the tool to the CAD framework. Both types of tool integration are illustrated by Figure 6.7.

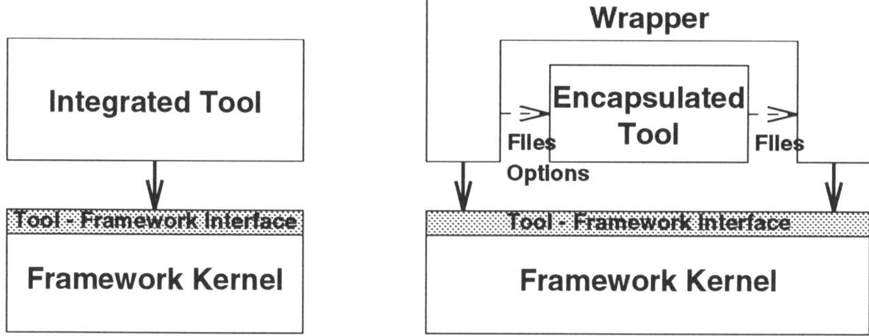

Figure 6.7   Integration versus encapsulation.

Upon integration, the tool itself is equipped with the proper calls to the DDMI functions (and design data handling functions), in order to interact with the framework while it is running. This is to take maximum advantage of the available framework services. Upon encapsulation, the interaction with the framework kernel is performed by the wrapper, which prepares the appropriate files, invokes the tool, and returns the result files to the framework. The wrapper must infer from the result files which operations the tool has performed during its run, and inform the framework of those operations. This post-mortem inference may be inherently cumbersome and inaccurate. Since our framework architecture is aimed at run-time interaction of tools with the framework kernel, we support both integration and encapsulation. The advantages of this approach have been stressed in section 6.2.

6.9 FRAMEWORK TOOLS

6.9.1 Introduction

In chapter 3 we concluded that the CAD framework consists of a framework
kernel and framework tools. We defined framework tools as *domain neutral
tools* aimed specifically at interaction of the end-user with the framework. In
our global framework model (Figure 3.4) we represented framework tools and
design tools as individual modules. This reflects some important properties
of tools in our framework architecture:

- Multiple tools can be operated concurrently by the end-user.

- Tools can be operated at different locations in the distributed hardware
  environment.

- Tools can be added or replaced on an individual basis.

Each running tool has its own internal operation and maintains an internal
state. When so desired, it interacts with the framework kernel to have functions performed. Also, one tool can order the execution of another tool. In
the next chapter we will see that the obvious choice upon implementation is
to map each running tool to a Unix process.

Framework tools make framework functions available to end-users. These
are functions to inform the end-user about the state of design, as maintained
by the framework kernel, as well as functions to operate on the state of
design. In addition, a variety of utility functions may be offered to end-
users. To give a more concrete idea, we present the following incomplete
list of topics:

- Browsing the structure and status of the design.

- Tool invocation.

- Project level operations (create project, remove project, etc).

- Design object level operations (remove, import, install, etc).

- System administration (registering users, tools, etc).

- Monitoring of system behavior.

- Backup and archive facility.

- Error log inspection.

- etc.

For these and other topics, framework tools may provide a great variety of functions. These functions typically call functions in the framework kernel. For example, the framework tools that provide the design object level operations to the end-user call the DDMI functions to perform the actual operations on the selected design objects. The browse facilities interact directly with the Meta Data Handling component, to access the meta data for presentation to the end-user.

6.9.2   Browsing and Tool Invocation

In the context of framework tools, we specifically address the issue of *designer assistance*. In our view, a CAD framework is much more than a platform on top of which a couple of tools can be tied together. In chapter 1 we already proclaimed that the CAD framework is to become the electronic assistant of the designer.

In our view the daily jog-trot of the design engineer looks as follows:

```
while  ( not finished )  {
    select data;
    select design tool;
    invoke design tool;
    operate design tool;
    organize data;
}
```

The CAD framework is to support the design engineer in executing these steps as efficiently as possible. Stage one in our approach has been to let design tools interact with the framework kernel, which organizes the design data and maintains a rich administration about the state of design. We now enter stage two, which is the exploitation of this administration for the benefit of the end-user; Framework tools are to reap what has been sown by the kernel.

One of the problems with today's design systems is the lack of support for the end-user to keep track of the state of his design. Typically, a design engineer must remember a lot of information about the design, in order to be able to effectively decide which operation to perform where. Retrieving useful information from the system often is a cumbersome activity. A design engineer typically spends a significant amount of time answering questions like:

- "What is my latest simulation result for this circuit and which stimuli were used to obtain it?"

- "Did I re-synthesize a gate-level description since I changed this HDL description, and if so, which one was derived from it?"

- "Where did I use the latch I designed last week?"

This situation is a possible source of design errors and adversely influences design productivity. This is particularly true if multiple design engineers are to cooperate in a design team.

A prominent step towards designer assistance is to provide facilities that allow the design engineer to *browse* in a highly convenient way through the administered state of design. These framework browsers must *present information* about the structure and status of the design in an attractive and comprehensible way to the end-user. They must offer convenient means to *navigate* through the available information, to explore the state of design.

The framework kernel maintains a variety of information about the design objects. See again the data schema derived in chapter 5 (Figure 5.22). In particular, the following types of information are of interest to the end-user:

- Version history.
- Hierarchical composition.
- Equivalence relationships.
- Design flow status.

Note that the data schema merely defines the structural semantics of this information and still leaves much freedom to choose the appropriate graphical representation in the framework browsers.

The hierarchical, equivalence and version derivation relationships may yield a highly interconnected web of design objects. Both the number of objects and the number of relationships may be large. The approach taken by Gedye [GK88] and the Nelsis CAD Framework [BvdW90] is to have distinct browsers for the different types of relationships in which design objects may participate. An individual browser presents graph-like pictures of design objects and their relationships of a particular type. Pruning techniques are employed to control the amount of information that is displayed. An example is the hierarchy browser of the Nelsis CAD Framework shown in Figure 6.8.

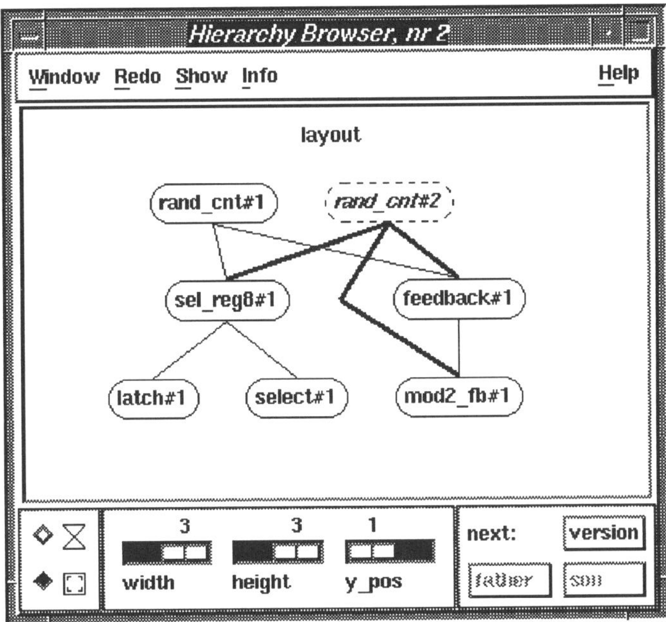

Figure 6.8 Hierarchy browser of the Nelsis CAD Framework displaying design objects and their hierarchical relationships.

A variety of modern user interface techniques can be employed to allow convenient navigation through the web of design objects. For example, in the Nelsis CAD Framework browsers can be activated on design objects via pop-up menus, and design objects can be passed from one browser to the other using an intuitive drag & drop facility.

To effectively support the design engineer in deciding which tool to run on which data, the framework must be able to show which tools have been run on which data and which tools can be run next. This brings us to the area of design flow management. As we described in section 5.12, the CAD framework knows the configured design flow and maintains for each design object the status with respect to this design flow. This information can also be presented graphically to the design engineer.

In the literature we see a variety of techniques to visualize design history, each one closely connected to the overall approach taken towards design flow management. For example, VOV [CNSVl90] can visualize a *design trace*, showing a historical record of tool runs and the data involved in these tool runs. VOV has no notion of a configured design flow.

The Hercules system [BD91] allows the design engineer to graphically derive a template *task tree* from a task schema which represents all possible task sequences (a task schema corresponds to a design flow). A task tree has a target entity (for example 'simulation output') as its root, and the tools and types of data required for producing this target entity as its leafs. The Hercules instance browser allows users to assign actual data items and perform forward and backward searches. Once data items have been assigned, a task tree can be executed.

The approach taken by van den Hamer et. al. [vdHT90] and the Nelsis CAD Framework [tBBvdW91, tBvdWB93] aims at the visual integration of the configured design flow ("what can be done") and the state of design ("what has been done") in a single graphical representation. The design flow is the template that can be *colored* with information about design objects, tool runs and their dependencies. This is termed *flow coloring* in [tBvdWB93]. A colored flow can visualize in an intuitive way the intricate relationships among tools and data. Design objects can be presented in the context of executed, enabled and disabled tools and related design objects. The principle of flow coloring provides the basis for a flow-based user interface, or *flow browser*, that permits the end-user to interact with a colored design flow. Such a flow browser naturally supports data selection, design history browsing, tool selection and tool invocation.

Figure 6.9 shows the flow browser of the Nelsis CAD Framework at work.

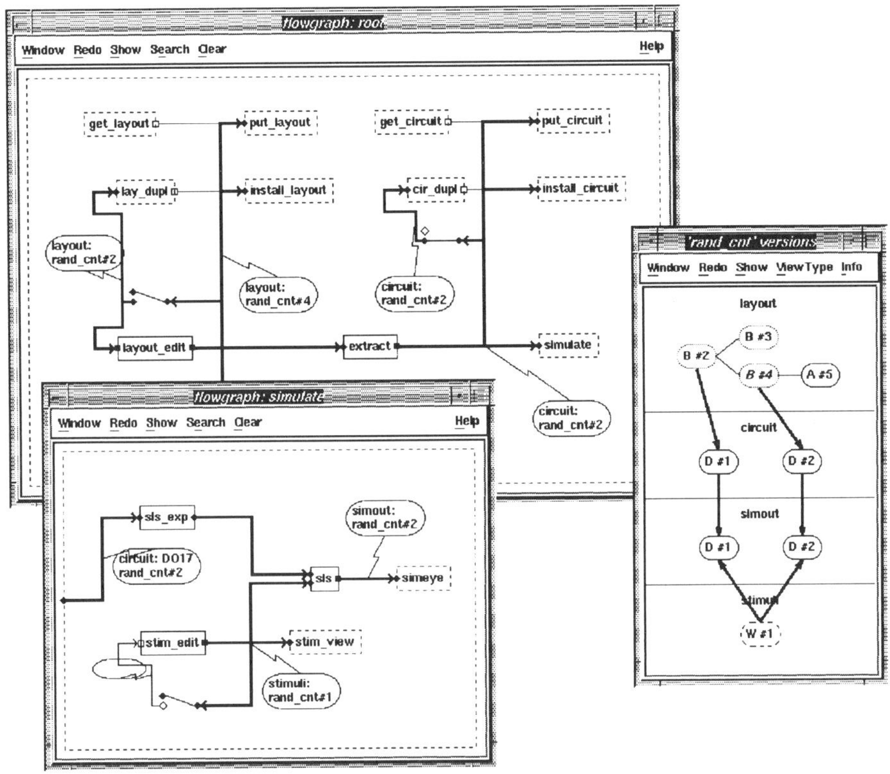

Figure 6.9 Flow browser of the Nelsis CAD Framework showing the colored 'root' flow graph and the expanded 'simulate' flow graph, while the version browser shows the versions of the 'rand_cnt' cell with their (horizontal) version derivation and (vertical) equivalence relationships.

The flow browser displays the configured hierarchical design flow. The 'simulate' flowgraph has been expanded and the version derivation history is shown simultaneously in the version browser. For the selected design objects, the flow browser shows which activities have been performed and which can be performed. At each instance in time, the flow browser can project only a small part of the actual state of design on the design flow. Browse functions permit the design history to be further explored. The flow browser fully cooperates with the other Nelsis browsers via the drag & drop facility. For example, a design object can be picked with the mouse from the version browser, to be dropped in the flow browser to have its design flow status displayed.

Once the design engineer has decided which tool to run on which data, the CAD framework must allow him to invoke the tool conveniently. For example, it may support argument specification and option selection for convenient command line building. The flow browser of the Nelsis CAD Framework allows a tool to be invoked simply by selecting it in the displayed design flow via a mouse-click. This is one of the advantages of using design flow as the basis for browsing the design data and the design history. Upon tool selection, a form pops up, showing the possible arguments and options of the selected tool (which have been configured by means of a TES file, according to the Tool Encapsulation Specification (TES) standard of CFI [CFI92c]). Where appropriate, argument values have already been assigned, based on the selected design objects in the flow browser. The design engineer can complete the form and run the tool. An example tool invocation form is displayed in Figure 6.10.

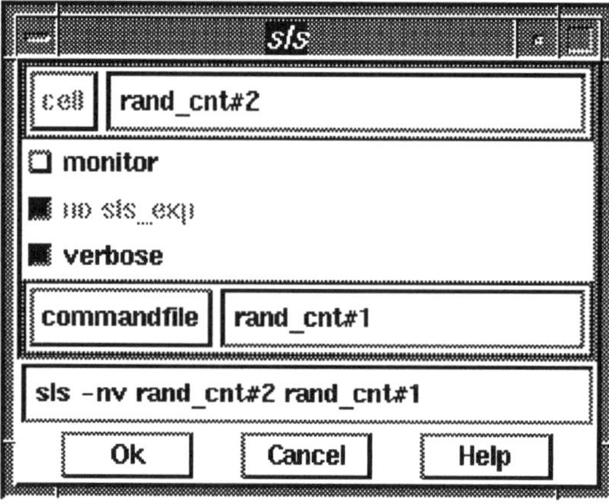

Figure 6.10 Example tool invocation form.

The individual browsers of the Nelsis CAD Framework have been integrated into a single framework tool [BvdW90]. This tool can be run in parallel to the actual design tools, which permits design engineers to switch between browsing and tool operation simply by moving their mouse. This tool also permits project selection, allowing the design engineer to move around in the logically distributed design environment.

We conclude that a properly integrated set of browse and tool invocation facilities can turn a CAD framework into the electronic assistant of the designer, provided it has a proper foundation for the administration of the state of design. The browse and tool invocation facilities can help the design engineer to efficiently perform the steps of data selection, design tool selection and design tool invocation. Due to increased observability, this process also becomes less error prone. As a result, the design engineer can effectively spend his time on what he likes to do most: design.

6.9.3 Notification

In the introduction of this section we remarked that "a running tool has its own internal operation and maintains an internal state". This poses a consistency problem if this state is required to correspond to the state of design as administered by the framework kernel. Multiple tools may be running simultaneously, each one manipulating the state of design. Data which is retrieved by a tool may be subject to change right after the locks on this data have been released.

This problem is particularly urgent for the browse tools. These tools typically take a *snapshot* of (part of) the state of design, as administered in the meta data, to present it in an attractive way to the end-user for a possibly longer time. They are not allowed to hold locks on the meta data while the information is being displayed, since access must be permitted for other tools. Some strategy must be adopted to let these browse tools refresh their snapshot when it is suspected to be out of date.

One strategy is to leave the initiative to the tools themselves. In this situation, the tools have to be active to see if their snapshot has to be refreshed. The alternative is to have the framework kernel provide some facility which informs the interested tools about updates on the meta data. We are in favor of the latter, and introduce a *notification service* in the framework kernel. The requirements for such a notification service are:

- It has to notify any interested tool when an update has been performed on the meta data.

- With the notification, it has to provide specific information about the update that has been performed. This allows a tool to decide whether its snapshot is still valid, and thereby prevents unnecessary actions.

- It must allow notifications to be *buffered* when many meta data updates are performed in a short time frame. This is to keep browsers from react-

ing when a new snapshot is known to be invalidated shortly afterwards
(we're talking subseconds here).

The last two requirements deal with performance as well as user convenience.
Satisfaction of the second requirement allows a browser to judge quickly
whether the update of the meta data will affect his 'picture'. Satisfaction of
the third requirement will prevent 'flashing' browsers that try to keep pace
with a multitude of meta data updates that are performed one after another.

The component architecture basically leaves us two choices for locating the
notification function. These are indicated by the dashed arrows in Figure
6.11. Note that the arrows are intended merely to identify the component
that sends the notifications, rather than representing a calling dependency.

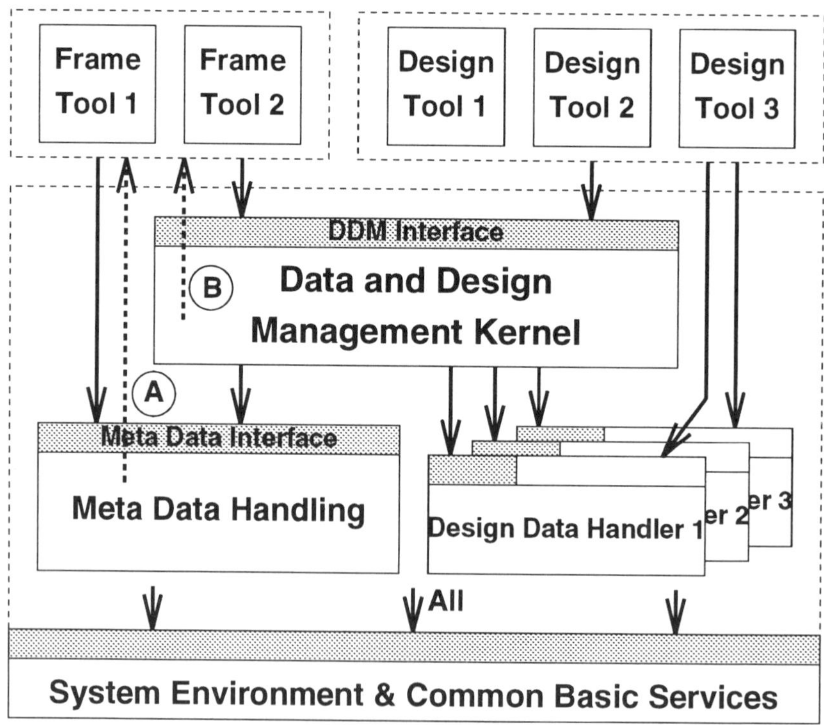

Figure 6.11 Alternative solutions A and B for incorporating a notification
service.

According to alternative A (see the dashed arrow tagged with an A in Figure 6.11), the Meta Data Handler sends the notifications. This is a *generic* facility. Without buffering, the Meta Data Handler can send a notification at the end of each meta data transaction. A notification may be accompanied by information about the object types involved in the meta data transaction as well as information about the instances that were manipulated. The Jessi-Common-Framework performs notification at the level of the data handler [JCF91b].

Alternative B is a more specific solution. The DDM Kernel is equipped to send typed notifications for the set of transaction types that it may perform. That is, a notification dictionary is agreed upon for a particular framework release. A notification may be accompanied by information about the design objects involved in a transaction, for example their names and view types, and auxiliary information such as access modes. According to the component architecture, all meta data manipulations for which notifications have to be sent, must be located in the DDM Kernel. Alternative B has been implemented in the Nelsis CAD Framework.

The advantage of alternative A is the genericness. It is elegant and facilitates framework evolution. The advantage of alternative B is that it provides more control over what to send when. For both alternatives, higher level components must be allowed to temporarily halt the sending of notifications when they are about to initiate multiple meta data updates in a short time frame. At the end of the update sequence they may order the buffered notifications to be sent out as one compound notification. When they fail to do so, the notification service must send out the buffer on its own initiative.

6.10   SUMMARY AND CONCLUSION

In this chapter we have defined the component architecture of the CAD framework. This architectural view addresses the key elements of the internal structure of the framework. The framework kernel has been decomposed into a number of coarse-grain framework components, with clear principles for the allocation of functions to components. For each component the major characteristics and principal functionality were described. The dependencies among components were identified, and interface definitions were presented.

We summarize the key steps in the definition of the component architecture as follows:

- We distinguish between a framework kernel and framework tools.

- We identify three major sub-components for the framework kernel:
  1. The System Environment and Common Basic Services component.
  2. The Data Handling component.
  3. The Data and Design Management Kernel (DDM Kernel).

- We refine the component architecture by decomposing the Data Handling component into a Meta Data Handler and a Design Data Handler. The advantage is increased modularity and flexibility. Procedures for maintaining meta data - design data consistency are defined.

- The major characteristics and principal functionality of a flexible high level Meta Data Handler are defined. This component is to become the *work horse* of the CAD framework.

- At the side of the Design Data Handler *openness* is pursued. An interface is defined that permits alternative data handlers to be incorporated.

- A variety of data and design management functions are offered by the DDM Kernel on top of the Data Handler. We do not discuss the detailed functionality, but demonstrate how these functions are realized as meta data transactions, with corresponding design object level operations on the Design Data Handler.

- A well-structured DDM Interface is defined based on a layered transaction schema that localizes the interaction between the tools and the DDM Kernel. Both encapsulation and tight integration are supported.

- We discuss the framework tools, with a focus on browse and tool invocation facilities, to turn our CAD framework into the *electronic assistant of the designer*.

While deriving the component architecture, we described were and how to handle such global issues as logical distribution, concurrency control, and access control. We conclude that a coarse partitioning of the framework kernel and a limited number of principles have allowed us to clarify how a great variety of issues is handled inside the framework. Starting from this component architecture we will define the implementation architecture in the next chapter.

7

THE IMPLEMENTATION ARCHITECTURE

7.1 INTRODUCTION

In this chapter we present the implementation architecture of the CAD framework. This architectural view provides additional detail on the internal structure of the framework. As distinct from the component architecture, which views the internals of the framework from a logical perspective, we will now describe these internals at the physical level. The design choices made at this level are decisive for obtaining maximum efficiency and optimal behavior with respect to physical distribution and multi-user support.

We will specifically aim at satisfaction of the following requirements:

- *Performance.* The framework must permit efficient interaction, both for tools and end-users.

- *Distributed access.* Design information must be accessible from all machines in the distributed computing environment.

- *Transparency of distribution.* Details of physical distribution must be hidden from tools and end-users.

- *Concurrent access.* Under control of the locking procedures, multiple tools must be allowed to operate concurrently on design information, even when running on different machines.

- *Scalability.* Framework capacity must be able to grow with the size of the design environment (design size, number of end-users, tools, machines, etc).

- *Configurable, flexible, extensible, modular, portable, maintainable.* We aim at a well-structured implementation that allows the framework to adjust and evolve.

The objective of the implementation architecture is to define how the CAD framework is implemented in terms of operating system primitives. As the implementation medium we take the Unix operating system, which is today the de facto standard for the more demanding engineering applications.

7.2 IMPLEMENTATION PRIMITIVES

We start with a description of the primitives in terms of which the implementation architecture is to be expressed. A computer-based system consists of a collection of programs, programming libraries and data files. When a program is actually running, we call it a *process*. A process is an address space, a single thread of control that executes within that address space, and associated system resources [CFI92a]. A process runs on a particular machine. Unix is a multi-tasking operating system: it can simultaneously execute multiple programs.

Processes are not forced to live in isolation. Interprocess communication (IPC) facilities permit processes to exchange or share data. This extends the ability to structure software systems as multi-program organizations. Unix offers several IPC facilities, each one aimed at a particular category of applications. The best IPC method for a given application depends on the structure of the communicating programs, the amount and kind of data that must be passed, the requirements for operation in a distributed computing environment, and the demanded performance.

In the simplest case, processes can communicate by writing to and reading information from *files*. The advantages of file IPC are unlimited capacity and multiple delivery. The classic Unix mechanism for IPC on a single machine is the *pipe*, which employs the basic byte-stream model used for file I/O. Via a pipe, a process may provide data for direct consumption by another concurrent process. *Named pipes*, or FIFOs, are identical to pipes in operation, but are not constrained to have the channel set up by a common ancestor. Facilities for memory-based IPC on a single machine are: *message queues*, *shared memory*, and *semaphores*. These facilities do not require common ancestry. In [Wat87] an "IPC use taxonomy" is presented for the above facilities. Four classes of IPC uses are identified: event trigger, state indicator, message exchange, and data transfer. For these classes the most appropriate IPC facilities are indicated.

We want our system to operate in a *distributed* computing environment. Figure 7.1 presents an intuitive view of such an environment, with machines, with or without disk(s), connected to a network.

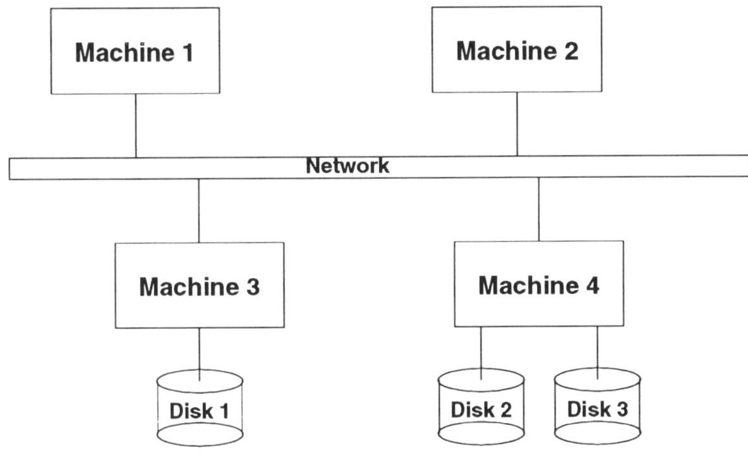

Figure 7.1   Intuitive view of a distributed computing environment.

Processes living on different machines may have a need to exchange or share data. Unix offers facilities for IPC between processes on (possibly) different machines according to the file I/O byte stream model. These transport facilities are known as *sockets* (from BSD) or *TLI* (from System V). They provide the basic service of reliable, end-to-end data transfer across the network.

The *Network File System* (NFS) is a facility for sharing files in a heterogeneous environment of machines connected to a network. NFS makes all disks available as needed. As a result, individual machines have access to all file-based information residing anywhere in the network.

A powerful model for IPC is the *remote procedure call* (RPC) model [BN84]. The basic idea of RPC is to extend the use of procedure calls to a distributed environment. With RPC, a *client* process can have a procedure executed by a *server* process, which may be running on a different machine. That is, a single thread of control logically winds through two processes. When a remote procedure is called, the client is suspended, a message containing the arguments is constructed and passed to the server process which executes the procedure. When the procedure finishes, the results are passed in a message back to the client process, and the client resumes as if the procedure had run locally. RPC is an attractive communication mechanism for distributed programs because it is simple, familiar, general, and can be implemented efficiently [Mul89].

An RPC server program implements one or more remote procedures; the procedures, their parameters and results are part of the specific program's *protocol specification*. An RPC service can be implemented on top of a basic transport facility such as the socket IPC facility. Some issues that arise in the design and use of an RPC service are:

- *Protocol specification.* Client and server programs must agree on a set of procedures and their interfaces. To permit framework evolution, a strategy for safe protocol evolution must be defined.

- *Heterogeneity.* Different kinds of machines may be attached to the network, employing different representations of data. Upon transfer of data between different computer architectures, conversion between representations must be performed.

- *Transparency.* To what extent do the *semantics* of an RPC match that of a local procedure call? First, an RPC is more susceptible to *failures*. Even if using a reliable transport such as TCP/IP, the client still needs to cope with server crashes. Achieving transparency also may be complicated by the fact that the procedure is executed in a different address space. Parameters must be passed *by value*. The more the operation of a procedure depends on local data structures, the less suited it becomes for implementation as a remote procedure.

The need to transfer data between the client and the server process, and the effect on performance, are crucial factors in deciding where and how to apply RPCs in distributed systems.

7.3   PROCESS ORGANIZATION

We have to define an organization of communicating processes for the execution of tool and framework functions. We start from the component architecture defined in chapter 6 (see again Figure 6.11). In this architectural view we have represented tools as individual modules. Multiple tools can be operated concurrently at the same or different locations in the distributed computing environment. The obvious design choice is to map each *running tool* to a Unix *process*. This fully matches with the paradigms of Unix and the X Window System in that multiple (graphical) applications can be run on one or more machines from multiple windows on a workstation.

As we described, the tools (or their wrappers) must interact with the framework kernel to obtain access to design information. Upon this interaction, the framework kernel maintains its meta data administration. We will study a number of process organizations to support the controlled sharing of meta data for multiple tool processes. Pictorially, we will represent processes by circles, data access via NFS by dashed lines, and RPC connections by solid lines.

A process organization of tool processes only is depicted in Figure 7.2.

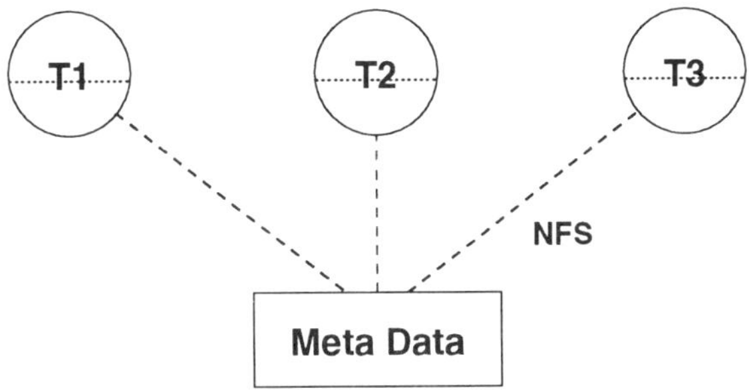

Figure 7.2   A process organization of tool processes only.

According to Figure 7.2, the framework kernel is linked to the tool programs (illustrated by the dotted lines in the tool processes). The meta data resides on disk, with access controlled via lock files. Upon a tool request, the kernel part of the tool process must obtain access to the meta data. NFS is used to provide distributed access. At the end of the request, the updated meta data administration must be made available for access by the other tool processes. It is obvious that serious performance problems will result from the extensive file I/O in the meta data access procedure.

A better process organization is obtained by having the tool processes communicate with a special *server process*. This is depicted in Figure 7.3.

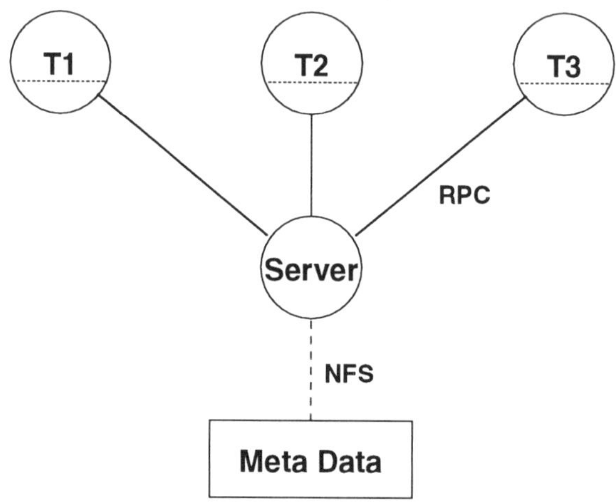

Figure 7.3 Multiple client - single server process organization.

Parts of the framework kernel are now running as a separate server process. Smaller parts are linked to the tool programs. Upon a tool request, the linked framework part issues one or more RPC requests to the server process. The RPC facility allows the processes to live on different machines. The server process controls the access to the meta data, and can be optimized for efficient meta data access. It maintains an up-to-date meta data administration on disk for recovery purposes. The major drawback of this process organization is that there is only one server process for the complete design environment. Hence, it does not *scale* with the number of tools being operated. The single server process is a potential *bottleneck*, or *hot-spot*, in the communication structure for meta data access. Moreover, if this server crashes, for whatever reason, all design activities are interrupted.

The bottleneck problem can be resolved by having multiple server processes serving the tool processes *in parallel*. This is depicted in Figure 7.4.

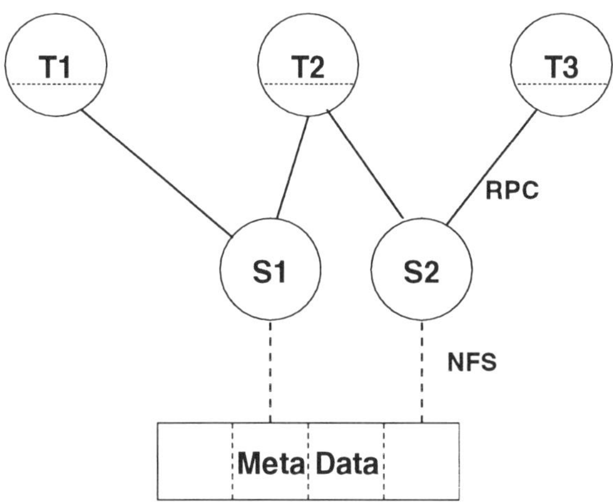

Figure 7.4   Multiple client - multiple server process organization.

This process organization is most effective if the individual servers can operate independently from each other. In section 6.5 we introduced the *logical distribution* of the meta data over project meta data repositories and a global meta data repository. Each meta data repository has its own data schema. A meta data transaction is issued against an individual meta data repository. The obvious design choice is to run a server process per meta data repository, and thereby exploit the parallelism implied by the logical distribution. Each server independently handles the clients for which accesses on the corresponding meta data repository have to be performed. In section 6.5 we required the possibility for higher level components to contact multiple meta data repositories in parallel. A client may, hence, be connected to multiple servers at the same time. See, for example, tool process T2 in Figure 7.4 which is connected to the servers S1 and S2. We will also refer to a server as a *project server*, with the global meta data repository being considered a special project. The principal task of a project server is to operate as *meta data server* for the corresponding project. Other tasks will be assigned later.

We can reflect the process organization in the intuitive view of the design environment that we presented in Figure 6.3. The result is presented in Figure 7.5. Project servers handle the access to the meta data repositories of the projects on which tools are active.

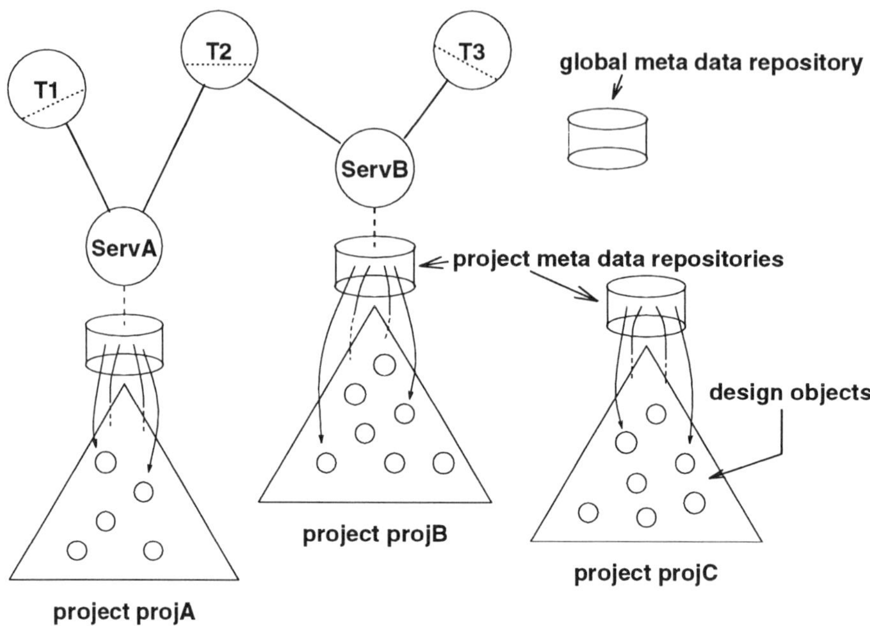

Figure 7.5   Project servers handle the access to the meta data repositories of
the projects on which tools are active (projA and projB).

The multiple client - multiple server process organization *scales well* with
the size of the design environment. The 'load' of a server has been reduced
to the number of tools operating *per project*. Hence, more project servers
will become active as more tools are being run in the design environment,
provided design activities are distributed well across different projects. The
multiple servers can effectively utilize the distributed computing power in
the workstation environment. Moreover, the system becomes more resilient
to server crashes. Upon a server crash, only the design activities in the
corresponding project are interrupted.

The relationships between the different object types involved in the process
organization are represented formally by the OTO-D data schema depicted
in Figure 7.6.

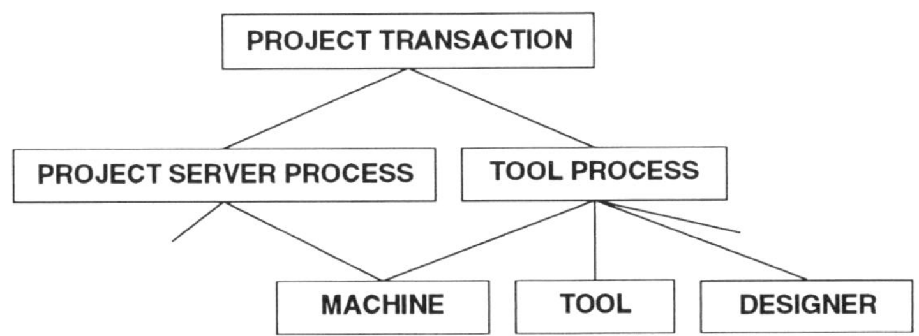

Figure 7.6 Data schema for the process organization.

A tool is run on a machine by a design engineer. A running tool may contact multiple servers, and a server can handle multiple tool processes simultaneously.

In section 6.4 we characterized the meta data as being *small* in size when compared to the volume of the corresponding 'raw' design data. *Queries* for small amounts of meta data are issued *frequently* by the higher level components. The relatively small size of the meta data permits a project server to handle the meta data of the corresponding project *incore* (i.e. in virtual memory). Meta data, structured according to the local data schema, is loaded into the address space of the project server when the project is activated. Subsequently, queries can be resolved *efficiently*. The project server is responsible for maintaining a correct persistent state of the meta data on disk. The server process, with its meta data, stays alive over multiple tool runs, until the project is de-activated. As a consequence of the survival of the server process, initialization time for successive tool runs is minimized. Having the meta data in the address space of the server process, perfectly matches with the use of RPCs for communication between the tool processes and the server processes. Query requests or higher level requests (to be defined later) that are sent to the server can efficiently access the incore meta data upon their execution.

The incore meta data handling by the project server implies a limit on the size of the meta data per project. The upper limit is about the size of the configured swap space of the operating system, which can be large (in the order of tens of Mbytes). Design projects having more meta data than this upper limit should be distributed over multiple smaller projects, which may be wise anyway. Since design data can be referenced across project boundaries, such distribution imposes no principal limitations.

In section 6.9 we introduced a *notification service* in the framework kernel.

This service permits a tool to register itself as an interested party which likes to be notified when updates are performed on the meta data of the project. In the process organization, the project server is the appropriate entity to handle the registration of interested tools as well as the distribution of notifications. We, therefore, assign the task of *notification handling* to the project server.

When a meta data update is performed on a project, the project server sends a notification to all interested parties. This is triggered either by the project server itself or by the framework part of the tool process that initiated the meta data update (see alternatives A and B in Figure 6.11). In the latter case the project server is informed by the tool process via an RPC. Special care must be taken at the client side by the interested parties, since notifications cause *asynchronous* events. For many graphical tools this can be tackled by incorporating the handling of notification events in the input event handling provided by the X Window System.

7.4 PROJECT IDENTIFICATION AND INITIALIZATION PROCEDURE

We do not require projects to reside on selected disks or file system partitions. Projects may reside anywhere in the distributed environment. The advantage is the *flexibility* for the end-user to freely select the location of his projects. Each project has a *project directory* somewhere in the distributed file system to hold project related files. A system-wide unique project identification is given by *hostname:abs_path*, where 'abs_path' is the absolute path from the root of the file system identified by 'hostname' to the location of the project directory. The abs_path is unique per host. We use the hostname:abs_path format internally for the unique identification of projects. This has the following consequences:

- When new projects are added to the design environment, no explicit check on uniqueness of identifications is required.

- There is no dependence upon a global catalog to map the project identification to a location on disk.

- For end-user convenience, the user interface may map the internal identifications to simple external identifications.

Given the project identification format, we must define how a project server gets started and how a tool process gets connected to the appropriate project server upon a ddmiOpenProject request (section 6.8). For this purpose we introduce a framework *daemon process*, which must be running on all machines where we want to run project servers. We summarize the initialization procedure as follows:

1. The framework part of the tool process, i.e. ddmiOpenProject, uses the project identification to access configuration files in the project directory (via NFS). From these files it retrieves the identification of the machine on which the corresponding project server should be running or must be started.

2. The tool process contacts the daemon on the identified machine to request (via an RPC) a project server handle for the identified project. The daemon maintains a list of the project servers that are active on his machine. If the project server is not yet running, it is started. The project server handle is returned to the tool process.

3. The tool process contacts the project server.

Note that proper measures must guarantee that at most one project server is active for a project. The file system mount tables of the operating system allow conversion of (relative) access paths to the unique project identification, and vice versa. Configuration files allow end-users to control where their servers are running. In a hardware environment of light-weight front-end machines and some larger server machines in the background, the project servers may be directed to the server machines. In an environment of many 'equal' workstations, the attractive default machine to direct project servers to, is the machine where the design engineer gets to work on the project.

7.5 RPC PROTOCOL SPECIFICATION

A tool process connects to the appropriate project server upon a ddmiOpenProject request (section 6.8). The connection remains until the tool informs the framework, via a ddmiCloseProject request, that operation on the project has been completed (connection mode service). In between, many RPCs may be issued by the framework part of the tool process to the project server. The project server maintains state information about its clients. An example is the tool registration for notifications, which we discussed above.

Upon the definition of the RPC protocol specification, we must decide which code goes into the server program and which code is linked to the tools. The criteria for code balancing are:

- Load on the server process.

- IPC message frequency and message sizes.

- Code size of the programs.

- Stability of the RPC protocol upon framework evolution.

- Need to re-link tools upon framework evolution.

- Ability to link customized versions of framework functions in the tool programs.

- Unix permissions of tool, server, and daemon processes.

We will not define in detail the RPC protocol specification. Instead, we will present some guidelines for the code balancing process, starting from the component architecture represented in Figure 6.11. The different levels at which we may define the RPC interface between a tool process and a server process, have been indicated in Figure 7.7.

An RPC interface may be defined at the highest possible level, that is, right under the DDM Interface and the Meta Data Interface (for the framework tools only). See the dotted line tagged with 'high' in Figure 7.7. This high level interface minimizes the code size of the tool programs, which contain only the code to set up the connection and a so called *stub* routine for each interface function. The stub routines send the input arguments in a request message, wait for a reply message and return the output arguments and function results. The tool programs do not have to be re-linked upon framework evolution, as long as the RPC protocol remains backward compatible. There are two major drawbacks for this type of RPC interface. First, all framework kernel code must execute in the server process, which can handle only one tool at a time. Second, there is no flexibility to link customized versions of certain framework functions in a tool program. This is a serious problem for our openness to alternative design data handling components. For reasons of simplicity and efficiency, we prefer to link the design object level interface functions (section 6.6) with parts of the DDM Kernel to the tool program, rather than further complicating the process organization.

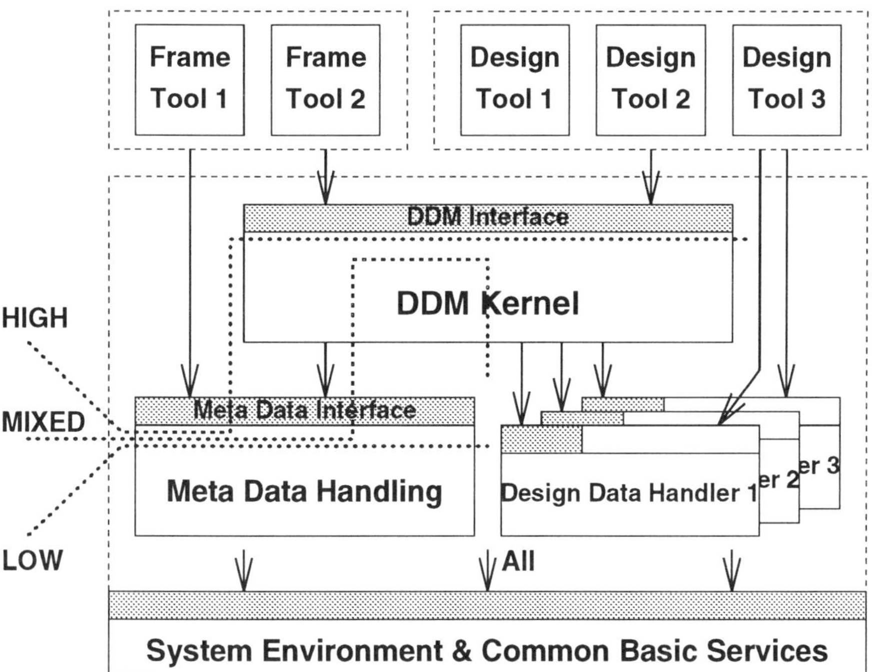

Figure 7.7 Different levels for the RPC interface between a tool process and a server process.

The other extreme is an RPC interface at the lowest possible level, that is, right under the Meta Data Interface (section 6.5). See the dotted line tagged with 'low' in Figure 7.7. The DDM Kernel gets linked completely to the tool programs, as well as the stub routines for the Meta Data Interface. The load of the project server is reduced, since part of the framework code executes in the tool processes. Note that this possibility follows from our decision to control concurrency at the level of the Meta Data Interface (section 6.5). The framework parts of the tool processes maintain their own state in local data structures, and contact the project server only upon meta data access. As we described in section 6.5, the Nelsis CAD Framework employs a declarative form of interaction with the Meta Data Handler. This has the advantage of a reduced communication overhead, as a complete query request in OTO-D DML syntax can be passed as one string to the project server. The query-level RPC protocol does not have to be changed when new framework services are incorporated in the DDM Kernel. This loose coupling at the Meta Data Interface facilitates framework evolution significantly. However, tool programs have to be re-linked upon framework evolution.

Some (sets of) functions in the DDM Kernel perform a large number of meta data accesses and do not share local data structures with other functions. These functions are excellent candidates for execution in the server process. The optimal solution, therefore, is a basic RPC interface at the low level of the Meta Data Interface, supplemented with RPCs for a selection of higher level functions. This allows the communication overhead to be reduced, and the server load and code size to be balanced. This mixed approach has been illustrated in Figure 7.7 by the dotted line tagged with 'mixed'. The Nelsis CAD Framework has a basic RPC interface at the level of the Meta Data Interface. In addition, some design flow management operations that query heavily on the state of design have been located in the server process. An additional advantage is that the data structures used by these functions do not have to be re-built upon each tool run.

According to the mixed approach for the RPC interface, the Design Data Interface is linked to the tool program. First, this simplifies interfacing to a foreign design data handler, as all access comes from the same process. Second, this approach implies minimal overhead in design data access. The Nelsis CAD Framework has a design data handler that maps each design object to a directory in the Unix file system. With this handler, all file access is performed directly from the tool process. Framework overhead in accessing 'raw' design data is thereby minimized.

7.6   APPLICATION OF SHARED MEMORY IPC

The communication overhead upon meta data access can be further reduced by the use of shared memory IPC. As has been described in [vdWSBD90], the project server can load the meta data into shared memory. *Local* tool processes, i.e. tools running on the same machine as the project server, can attach the shared memory to their address space. These tool processes may then access meta data directly in their address space rather than via RPCs. Besides increased efficiency there is increased concurrency in performing non-conflicting meta data accesses. A drawback is the increased complexity of the software, since low level synchronization must be performed via semaphores. Another problem is the limited size of the shared memory segment (in the order of several Mbytes). Further, the code size of the tool programs increases, as they have to carry their own Meta Data Handler. A tool process must be able to switch dynamically between local operation on shared memory and RPC-based meta data access. The gain in efficiency (i.e. the communication overhead) depends on the meta data access frequency and the size of the query results (rather than the size of the meta data repository). For some query intensive browse applications, a decrease in response time of about 50% was observed for local query execution. For other applications the efficiency gains were smaller.

7.7   PERFORMANCE

It is difficult to light on one single criterion by which the performance of a CAD framework can be judged. In our requirements (section 2.3) we stated that the key goal is overall optimization of design efficiency. This relates to run-time performance of individual framework services as well as to increased framework functionality that helps the end-user to work more effectively. Local inefficiencies are acceptable if they are instrumental in global optimization of design efficiency.

The Nelsis CAD Framework has been the vehicle for validating many of the ideas presented in this book. While evolving from one release to the next, it gained functionality to become an effective assistant of the design engineer. The work horse of the Nelsis CAD Framework is the Meta Data Handling component. The framework spends most of its computation time in this component. This is no surprise if we consider that many data and design management functions are structured as meta data transactions (section 6.7).

The two major components of the computation time spent on meta data handling are query parsing, as a consequence of the declarative form of interaction with the Meta Data Handler, and query execution. Query parsing is independent of the size of the meta data repository. Query execution time increases linearly with the size of the meta data repository. We remark that in this respect the semantic approach is superior to the relational data model: almost all relational operations are of squared order [tB88, tB91]. We also remark that our Meta Data Handling component is a straightforward implementation of the semantic data model that leaves much room for further optimization by the application of more advanced search techniques. This has not yet been given a high priority in the course of the Nelsis developments, since performance was felt to be satisfactory.

Two measurable aspects of the run-time performance are:

- the performance degradation of the design tool operations caused by the framework.

- the response of the interactive framework tools.

The performance degradation of the design tool operations observed upon the introduction of extensive data and design management facilities in the Nelsis CAD Framework, varies with the type of tool and the design size (in terms of the number of design objects and relationships between design objects). Two key operations are CheckOut and CheckIn, which bracket the design transactions initiated by the design tools. These operations perform many of the data and design management functions. For most tools they are the predominant consumers of framework CPU time. For medium-sized designs the CPU time spent on a CheckOut operation is in the range of 20 - 100 msec. on a SUN4 Sparcstation 2. For a CheckIn operation the CPU time spent is in the range of 10 - 80 msec. The large variations are caused by the conditional activation of data and design management functions. For instance, one design transaction may be a simple ReadOnly access, while the other has to create a new version, possibly overwriting an older version. For many design tools the relative performance degradation is in between 5% and 25%, when compared to operation on a single user system without advanced data and design management services. A SPICE simulation which accesses a single netlist description to simulate it extensively, endures only a very small performance degradation, since it performs very little interaction with the framework. An analysis tool which traverses the design hierarchy extensively, may endure a significant performance degradation. However, such tools may also take advantage of the framework services. For example, the services for consistency maintenance in hierarchical multi-view designs allow analysis tools to operate incrementally, leaving the administration of the correct state of design to the CAD framework. Upon iterative design activities this may yield significant savings of computation time. Also note that the concurrency control and distribution facilities allow the design engineer to fully exploit the distributed computing facilities.

In the Nelsis CAD Framework the response of interactive browse tools typically is instantaneous for medium-sized meta data repositories. The convenient browse and tool invocation facilities support the design engineer in efficiently selecting data and invoking tools. This helps him to be more productive.

7.8   SUMMARY AND CONCLUSION

In this chapter we have defined the implementation architecture of the CAD framework. This architectural view describes how the CAD framework is implemented in terms of Unix primitives. We have defined a multiple client - multiple server process organization, based on the logical distribution of the meta data. This process organization permits fast incore meta data handling, and scales well with the size of the design environment. Design information may be accessed from any machine on the network, and the distributed computing power can be utilized effectively. The RPC interface between the tool processes and the server processes is best located at the level of the Meta Data Interface (section 6.5) with optional extensions for a selection of DDM Kernel functions that query heavily on the meta data. Application of shared memory IPC may further increase efficiency and concurrency for local tools. Physical distribution is transparent to the tools, which simply initiate their requests via the framework interfaces defined in chapter 6.

We remark that the design choices made at the implementation level directly relate to high-level design choices made earlier in the framework development. Crucial factors are the meta data - design data distinction and the logical distribution of the meta data. From our experience we conclude that the proposed implementation architecture indeed permits the implementation of powerful *and* efficient data and design management facilities. These facilities outpace by far CAD frameworks that still rely on file based meta data handling.

8

CONCLUSION

We have given an in depth disquisition on the design and construction of CAD frameworks. Starting from the definition of principal framework requirements, we have systematically presented the architecture of a CAD framework by means of a global framework model and three more specific framework views. Key principles were defined to direct design choices and well-defined primitives were used to describe the framework views. The architecture describes the principal functions and global structure of the framework, as well as the principal mechanics of its inner structure.

A key characteristic of the framework architecture is the principal role allotted to the framework kernel. The framework kernel keeps track of the state of design by monitoring the data accesses performed by the tools. The knowledge of the state of design is exploited to enforce constraints on the design process and to assist the end-user in organizing the design information and managing the design process. This helps him to be more productive.

We distinguished between meta data and 'raw' design data contained in design objects. The meta data represents the information about the state of design maintained by the CAD framework. We have defined the structure of this information and described how it is handled inside the framework. We have illustrated how data and design management functions are incorporated. At the side of the design data we have pursued maximum openness. The resulting CAD framework allows different kinds of design tools to be integrated conveniently and operated effectively in a single environment.

Openness of the CAD framework was pursued by avoiding incorporation of features of a particular tool set or design representation. We focused on the handling of meta data and provided openness to a variety of storage regimes for the 'raw' design data. By supporting different levels of tool integration, the framework permits phased incorporation of design tools. It supports the full integration spectrum, from encapsulation to tight integration. Openness is further increased by the universal DDM Interface, which hides many framework specific details from the tools.

A number of design choices have been key to achieving good run-time performance. The coarse-grain operation at the design object level allows the framework to focus on efficient handling of the relatively small amount of meta data. The logical distribution of the meta data permits efficient operation in a physically distributed computing environment. Since framework intervention in handling detailed design data is minimized, design tools are not hampered in efficiently accessing their design descriptions.

The Nelsis CAD Framework has been the vehicle for validating many of the ideas presented in this book. This framework was evaluated extensively by different partners in the Jessi-Common-Frame project. They agreed unanimously that it demonstrated a superior run-time performance in combination with powerful data and design management services [JCF91a]. Up to the present a variety of (prototype) application environments has been built on top of the Nelsis CAD Framework. Some examples are:

- The Nelsis IC Design System [Dew86].

  This is an integrated CAD system for the design of integrated circuits. It offers an extensive tool suite for design tasks such as layout design and verification, circuit simulation, and place & route.

- The HiFi Design System [vdH92].

  This is a design system aimed specifically at the design of digital signal processor arrays. It is under continuing development at Delft University of Technology. Both encapsulation and tight integration are used to incorporate HiFi design tools.

- Analog Simulation Environment.

  At Philips Research Laboratories an analog simulation environment was constructed by integrating a set of tools on top of the Nelsis CAD Framework. The system includes tools for schematic editing, waveform editing, netlist expansion, simulation, and waveform display. The effort required for this exercise was reasonable (a few man weeks) and the resulting system demonstrated good run-time performance [JCF91a].

- Cathedral 2nd Synthesis System.

  At IMEC VZW, Belgium, the tool suite of the Cathedral 2nd high-level synthesis system [LCG+90] was integrated into the Nelsis CAD Framework. This exercise was performed in a few weeks using encapsulation techniques. The resulting system organizes all design information and allows the end-user to control his design activities from the flow-based graphical user interface.

The CAD framework has also been used successfully outside the domain of electronic design. Example applications were image generating simulations for X-ray equipment design and simulation and analysis for television tube design. In these cases too, the CAD framework was successful in turning collections of tools into effective user-friendly environments.

These practical experiences strengthen our conclusion that the presented architecture indeed provides the basis for an effective CAD framework that is open, efficient, and user friendly. Its data and design management functions support the execution of the design process and its graphical user interface permits convenient interaction of the end-user with the integrated environment. All in all, this makes the CAD framework the electronic assistant of the designer.

From our experience in the development and application of the Nelsis CAD Framework and our involvement in the Jessi-Common-Frame project we conclude that:

- An incremental approach is to be adopted to the design and implementation of a CAD framework, as is also stressed in [BHNS92].

- A layered system architecture is to be adopted, with a good allocation of responsibilities to the different layers.

- The data and design management functions must be conceptually harmonized.

- The system must be architected by a small number of people that are involved for a longer period of time.

- These people have to talk a proper data modeling language fluently.

- The experience of having actual tool suites integrated on top of the CAD framework is indispensable.

We believe that the topic of CAD frameworks can by no far be considered a "solved problem" and that there is still a lot of research to be performed. There is a great need for a formalization of data and design management issues. For example, the topic of change propagation in hierarchical multiview designs still seems to raise a lot of confusion. Approaches to design flow management need a more formal basis. The framework community needs a generally accepted understanding for such recurring topics, in order to provide a basis for further progress. We hope that the principles and architecture presented in this book help to establish such an understanding.

A lot of work is to be done in standardization. Application of framework technology is seriously hampered by the lack of (adherence to) standards. Further, we believe that new framework topics will arise. At present the areas of design methodology management, design decision support and project management leave much space for further exploration.

BIBLIOGRAPHY

[AM84]   H. Afsarmanesh and D. McLeod. A framework for semantic database models. In *Proc. NYU Symposium on New Directions for Database Systems*, New York, May 1984.

[BD89]   M. Bushnell and S.W. Director. Automated design tool execution in the ulysses design environment. *IEEE Trans. on CAD of Integrated Circuits and Systems*, CAD-8(3):279–287, March 1989.

[BD91]   J.B. Brockman and S.W. Director. The hercules cad task management system. In *Proc. IEEE ICCAD - 91*, pages 254–257, 1991.

[BG86]   L. Bic and J.P. Gilbert. Learning from ai: New trends in database technology. *IEEE Computer Magazine*, 19(3):44–54, March 1986.

[BG87]   J. Brouwers and M. Gray. Integrating the electronic design process. *VLSI Systems Design*, pages 38–47, June 1987.

[BHNS92]   T.J. Barnes, D. Harrison, A.R. Newton, and R.L. Spickelmier. *Electronic CAD Frameworks*. Kluwer Academic Publishers, 1992.

[BK85]   D.S. Batory and Won Kim. Modeling concepts for vlsi cad objects. *ACM Trans. on Database Systems*, 10(3):322–346, Sept 1985.

[BK92]   M. Brielmann and E. Kupitz. Representing the hardware design process by a common data schema. In *Proc. EURO-DAC 92*, pages 564–569, Hamburg, Germany, Sept 1992.

[BKK85]   F. Bancilhon, W. Kim, and H.F. Korth. Transactions and concurrency control in cad databases. In *Proc. IEEE ICCD '85*, pages 86–89, 1985.

[BKL+90]   F. Bretschneider, C. Kopf, H. Lagger, A. Hsu, and E. Wei. Knowledge based design flow management. In *Proc. ICCAD - 90*, pages 350–353, 1990.

[BN84]   A.D. Birrell and B.J. Nelson. Implementing remote procedure calls. *ACM Trans. on Computer Systems*, 2(1):39–59, Feb 1984.

[BP86]   M. Benayoune and P.E. Preece. Methodology for the design of databases for engineering applications. *Computer-Aided Design*, 18(5):257–262, June 1986.

[BT90]   J. Bhat and F. Taku. A seven-layer model of framework functionality. *Electronic Engineering*, pages 67–73, Sept 1990.

[BtBvdW92]   P. Bingley, K.O. ten Bosch, and P. van der Wolf. Incorporating design flow management in a framework based cad system. In *Proc. IEEE/ACM International Conference on CAD - 92*, pages 538–545, Santa Clara, Nov 1992.

[Buc84]   A.P. Buchmann. Current trends in cad databases. *Computer-Aided Design*, 16(3):123–126, May 1984.

[BvdW90]   P. Bingley and P. van der Wolf. A design platform for the nelsis cad framework. In *Proc. 27th ACM/IEEE Design Automation Conference*, pages 146–149, Orlando, June 1990.

[BZ81]   M.L. Brodie and S.N. Zilles, editors. *Proc. of the Workshop on Data Abstraction, Databases and Conceptual Modelling at Pingree Park, June 23-26 1980*, Jan 1981. Sigplan Notices, 16(1).

[Cad90]   Framework technology for the 1990s. Cadence Design Systems, 1990. Cadence Marketing Materials.

[CFHL86]   K.C. Chu, J.P. Fishburn, P. Honeyman, and Y.E. Lien. A database-driven vlsi design system. *IEEE Trans. on CAD of Integrated Circuits and Systems*, CAD-5(1):180–187, Jan 1986.

[CFI90a]   CFI Architecture Technical Subcommittee. *Suggested Framework Problem Statement*, 1990. CAD Framework Initiative.

[CFI90b]   CFI Architecture Technical Subcommittee. *CAD Framework Users, Goals, and Objectives, Version 0.91*, Aug 1990. CAD Framework Initiative.

[CFI91a]   CFI Architecture Tiger Team. *Framework Architecture Reference, Version .81*, Oct 1991. CAD Framework Initiative.

[CFI91b]   CFI Architecture Tiger Team. *Framework Views Provide Architectural Insight*, Fall 1991. The Initiative, pages 8-12.

[CFI92a]   CAD Framework Initiative, Inc. *Computing Environment Services, Version 1.0.0*, 1992.

[CFI92b]   CAD Framework Initiative, Inc. *Design Representation Programming Interface: Electrical Connectivity, Version 1.0.0*, 1992.

[CFI92c]   CAD Framework Initiative, Inc. *Tool Encapsulation Specification, Version 1.0.0*, 1992.

[CFI93]   CFI Architecture Working Group. *Framework Architecture Reference, Version 1.2*, June 1993. CAD Framework Initiative.

[Che76]   P.P. Chen. The entity-relationship model: Toward a unified view of data. *ACM Trans. on Database Systems*, 1(1):9–36, March 1976.

[CK88]   H.T. Chou and W. Kim. Versions and change notification in an object-oriented database system. In *Proc. 25th ACM/IEEE Design Automation Conference*, pages 275–281, Anaheim, June 1988.

[CNSV190]   A. Casotto, A.R. Newton, and A. Sangiovanni-Vincentelli. Design management based on design traces. In *Proc. 27th ACM/IEEE Design Automation Conference*, pages 136–141, 1990.

[Dat86]   C.J. Date. *An Introduction to Database Systems, Fourth Edition*, volume I. Addison-Wesley Systems Programming Series, 1986.

[DD89]   J. Daniell and S.W. Director. An object oriented approach to cad tool control within a design framework. In *Proc. 26th ACM/IEEE Design Automation Conference*, pages 197–202, 1989.

[DEC92]   Digital Equipment Corp. *Data Management Yields a High Return on Investment*, June 1992. Electronic Design, pages 88G-88N.

[Dew86]   P. Dewilde, editor. *The Integrated Circuit Design Book: Papers on VLSI Design Methodology from the ICD-NELSIS Project.* Delft University Press, Delft, The Netherlands, 1986.

[DFW82]   M. Deering, J. Faletti, and R. Wilensky. Using the pearl ai package. Technical report, University of California, Berkeley, Feb 1982.

[dG93]   Aart de Geus. 1.000.000 gate asic? not with present eda tools! In *Proc. IEEE International ASIC Conference and Exhibit*, Sept 1993.

[dSSV90]   P. R. dos Santos, H. Sarmento, and L. Vidigal. Ghost / spook, user interface and process management in the pace framework. In *Proc. European Design Automation Conference*, pages 501–505, Glasgow, Scotland, March 1990.

[EDI90]   Electronic Industries Association. *EDIF - Electronic Design Interchange Format, Version 2 0 0*, second edition, 1990.

[EGLT76]   K.P. Eswaran, J.N. Gray, R.A. Lorie, and I.L. Traiger. The notions of consistency and predicate locks in a database system. *Communications of the ACM*, 19(11):624–633, Nov 1976.

[Exp91]   International Standards Organization. *EXPRESS Language Reference Manual*, March 1991. ISO TC184/SC4/WG5 Document N14.

[FFHe91]   W. Fox, J. Friedrich, R. Hopp, and T. Kathofer et.al. The architecture of the object management system within the cadlab framework. In F.J. Rammig and R. Waxman, editors, *Proc. IFIP WG 10.2 Workshop on Electronic Design Automation Frameworks, Nov 1990*, pages 141–154. North-Holland, Charlottesville, VA, USA, 1991.

[GE87]   A.J. Gadient and J.L. Ebel. Rationale for and organization of the engineering information system program. In *Proc. 24th ACM/IEEE Design Automation Conference*, pages 764–769, 1987.

[GK88]   D. Gedye and R.H. Katz. Browsing the chip design database. In *Proc. 25th ACM/IEEE Design Automation Conference*, pages 269–274, Anaheim, June 1988.

[Goe85]   R. Goering. Design tool vendors move toward integrated systems. *Computer Design*, pages 103–122, June 1985.

[GS88]   K. Goldman and T. Stout. A design automation environment. *VLSI Systems Design*, pages 46–49, June 1988.

[Har84]   M. Hardwick. Extending the relational database data model for design applications. In *Proc. 21st ACM/IEEE Design Automation Conference*, pages 110–116, 1984.

[Hay81]   M.N. Haynie. The relational/network hybrid data model for design automation databases. In *Proc. 18th ACM/IEEE Design Automation Conference*, pages 646–652, 1981.

[Hay83]   M.N. Haynie. Tutorial: The relational data model for design automation. In *Proc. 20th ACM/IEEE Design Automation Conference*, pages 599–607, 1983.

[HJR92]   W. Heijenga, U. Jasnoch, and E. Radeke. Dadamo - a conceptual data model for electronic design applications. In *Proc. European Design Automation Conference*, pages 394–398, Brussels, March 1992.

[HMSN86]   D.S. Harrison, P. Moore, R.L. Spickelmier, and A.R. Newton. Data management and graphics editing in the berkeley design environment. In *Proc. IEEE ICCAD-86*, pages 24–27, 1986.

[HNCL84]   L. Hollaar, B. Nelson, T. Carter, and R.A. Lorie. The structure and operation of a relational database system in a cell-oriented integrated circuit design system. In *Proc.*

21st ACM/IEEE Design Automation Conference, pages 117–125, 1984.

[HNSB90]   D.S. Harrison, A.R. Newton, R.L. Spickelmier, and T.J. Barnes. Electronic cad frameworks. *Proceedings of the IEEE*, 78(2):393–417, Feb 1990.

[HPC93]   A.R. Hurson, S.H. Pakzad, and J.B. Cheng. Object-oriented database management systems: Evolution and performance issues. *IEEE Computer Magazine*, pages 48–60, Feb 1993.

[HS87]   M. Hardwick and D.L. Spooner. Comparison of some data models for engineering objects. *IEEE CG&A*, pages 56–66, March 1987.

[HY85]   M. Hardwick and N. Yakoob. Using a database system and unix to author cad applications. In *Proc. IEEE ICCAD - 85*, pages 53–55, 1985.

[Jac83]   M. Jackson. *System Development*. Prentice Hall, Englewood Cliffs, N.J., 1983.

[Jan86]   A. Di Janni. A monitor for complex cad systems. In *Proc. 23rd ACM/IEEE Design Automation Conference*, pages 145–151, 1986.

[JCF90]   JCF Task Force Architecture. *Jessi-Common-Framework Overall Requirements*, Nov 1990. Jessi-Common-Frame project (Esprit 5082 / 7364).

[JCF91a]   JCF Evaluation Sub-Project. *Start-up Phase Report (D4.1)*, June 1991. Jessi-Common-Frame project (Esprit 5082 / 7364).

[JCF91b]   JCF Task Force Architecture. *Jessi-Common-Framework Architecture Specification*, June 1991. Jessi-Common-Frame project (Esprit 5082 / 7364).

[JCF92]   JCF SP2 Task Forces. *Jessi-Common-Framework Glossary SP2*, Feb 1992. Jessi-Common-Frame project (Esprit 5082 / 7364).

[JD92]   M.F. Jacome and S.W. Director. Design process management for cad frameworks. In *Proc. 29th ACM/IEEE Design Automation Conference*, pages 500–505, 1992.

[JLL86]   C. Jullien, A. Leblond, and J. Lecourvoisier. A database interface for an integrated cad system. In *Proc. 23rd ACM/IEEE Design Automation Conference*, pages 760–767, 1986.

[Joh89]   W. Johnson. Bringing design management to the open environment. *High Performance Systems*, pages 66–70, June 1989.

[KAC86]   R.H. Katz, M. Anwarrudin, and E. Chang. A version server for computer-aided design data. In *Proc. 23rd ACM/IEEE Design Automation Conference*, pages 27–33, 1986.

[Kal85]   Y.E. Kalay. A database management approach to cad/cam systems integration. In *Proc. 22nd ACM/IEEE Design Automation Conference*, pages 111–116, June 1985.

[Kat82]   R.H. Katz. A database approach for managing vlsi design data. In *Proc. 19th ACM/IEEE Design Automation Conference*, pages 274–282, Las Vegas, Nv., June 1982.

[Kat83]   R.H. Katz. Managing the chip design database. *IEEE Computer Magazine*, 16(12):26–35, Dec 1983.

[Kat85a]   R.H. Katz. *Information Management for Engineering Design*. Springer-Verlag, Berlin, 1985.

[Kat85b]   R.H. Katz. Computer-aided design databases. *IEEE Design and Test*, pages 70–74, February, 1985.

[Kat86]   R.H. Katz. Design database. In S. Goto, editor, *Advances in CAD for VLSI*, volume 6, pages 501–525. Elsevier Science Publishers B.V. (North-Holland), 1986.

[Kat90]   R.H. Katz. Toward a unified framework for version modeling in engineering databases. *ACM Computing Surveys*, 22(4):375–408, December 1990.

[KBC+87]   R.H. Katz, R. Bhateja, E. E-Li Chang, D. Gedye, and V. Trijanto. Design version management. *IEEE Design & Test Magazine*, pages 12–22, Feb 1987.

[KGMB94]   S. Kleinfeldt, M. Guiney, J. Miller, and M. Barnes. Design methodology management. *Proceedings of the IEEE*, 82(2):231–250, Feb 1994.

[Knu81]   D. Knuth. *The Art of Computer Programming*, volume 2. Addison-Wesley Publ. Company, 1981.

[KW84]   R.H. Katz and S. Weiss. Design transaction management. In *Proc. 21st ACM/IEEE Design Automation Conference*, pages 692–693, 1984.

[LCG+90]   D. Lanneer, F. Catthoor, G. Goossens, M. Pauwels, J. van Meerbergen, and H. De Man. Open-ended system for high-level synthesis of flexible signal processors. In *Proc. European Design Automation Conference*, pages 272–276, Glasgow, Scotland, March 1990.

[LJ92]   D.C. Liebisch and A. Jain. Jessi-common-framework design management – the means to configuration and execution of the design process. In *Proc. EURO-DAC 92*, pages 552–557, Hamburg, Germany, Sept 1992.

[LP83]   R. Lorie and W. Plouffe. Complex objects and their use in design transactions. *Proc. Databases for Engineering Applications, ACM Database Week*, pages 115–121, May 1983.

[Lyn84]   P. Lyngbaek. *Information Modeling and Sharing in Highly Autonomous Database Systems*. PhD thesis, Univ. of So. California, Los Angeles, August 1984.

[Mal92]   L. Maliniak. Cad frameworks ride a rough road to success. *Electronic Design*, pages 36–40, August 1992.

[Man92]   H. De Man. Design technology research for the nineties: More of the same ? In *Proc. EURO-DAC 92*, pages 592–596, Hamburg, Germany, Sept 1992.

[May87]   M. Mayotte. An engineering design management system. *VLSI Systems Design*, pages 30–39, Jan 1987.

[MGSW89]   J. Miller, K. Gröning, G. Schulz, and C. White. The object-oriented integration methodology of the cadlab work station design environment. In *Proc. 26th ACM/IEEE Design Automation Conference*, pages 807–810, Las Vegas, June 1989.

[Mul89]   S. Mullender. *Distributed Systems*. Addison Wesley, Reading, Mass, 1989.

[Neu88]   V.E. Neufeldt, editor. *Webster's New World Dictionary of American English, 3rd college edition.* Webster's New World Dictionaries, 1988.

[NPSVtS81]   A.R. Newton, D.O. Pederson, A.L. Sangiovanni-Vincentelli, and C.H. Sequin. Design aids for vlsi: The berkeley perspective. *IEEE Trans. on Circuits and Systems*, CAS-28(7):666–679, July 1981.

[NSV86]   A.R. Newton and A.L. Sangiovanni-Vincentelli. Computer-aided design for vlsi circuits. *IEEE Computer Magazine*, pages 38–60, April 1986.

[Obj91]   Objectivity/db engineering database management system. Objectivity, Inc., 1991. Objectivity Marketing Materials.

[OTC93]   C. Oussalah, G. Talens, and M.F. Colinas. Concepts and methods for version modeling. In *Proc. EURO-DAC 93*, pages 332–337, Hamburg, Germany, Sept 1993.

[PM88]   J. Peckham and F. Maryanski. Semantic data models. *ACM Computing Surveys*, 20(3):153–189, September 1988.

[PT88]   W.D. Potter and R.P. Trueblood. Traditional, semantic, and hyper-semantic approaches to data modeling. *IEEE Computer Magazine*, pages 53–63, June 1988.

[RBJ81]   K.A. Roberts, T.E. Baker, and D.H. Jerome. A vertically organized computer-aided design data base. In *Proc. 18th ACM/IEEE Design Automation Conference*, pages 595–602, 1981.

[RF92]   M. Rumsey and C. Farquhar. Unifying tool, data and process flow management. In *Proc. EURO-DAC 92*, pages 500–505, Hamburg, Germany, Sept 1992.

[RRR$^+$88] S. Rehm, T. Raupp, M. Ranft, R. Laengle, M. Haertig, W. Gotthard, K.R. Dittrich, and K. Abramowicz. Support for design processes in a structurally object-oriented database system. In K.R. Dittrich, editor, *Proc. 2nd Intern Workshop on Object-Oriented Database Systems*, pages 80–97. Springer Berlin, 1988. ISBN 3-540-50345-5.

[SA86]   S.M. Staley and D.C. Anderson. Functional specification for cad databases. *Computer-Aided Design*, 18(3):132–138, April 1986.

[SBD+90]   G.W. Sloof, P. Bingley, P. Dewilde, T.G.R. van Leuken, and P. van der Wolf. Design data management in a distributed hardware environment. In *Proc. European Design Automation Conference*, pages 34–38, Glasgow, Scotland, March 1990.

[SBD93]   P.R. Sutton, J.B. Brockman, and S.W. Director. Design management using dynamically defined flows. In *Proc. 30th ACM/IEEE Design Automation Conference*, pages 648–653, 1993.

[SdS91]   H. Sarmento and P. R. dos Santos. Pace - a framework for electronic design automation. In F.J. Rammig and R. Waxman, editors, *Proc. IFIP WG 10.2 Workshop on Electronic Design Automation Frameworks, Nov 1990*, pages 85–97. North-Holland, Charlottesville, VA, USA, 1991.

[SGKN89]   M. Silva, D. Gedye, R.H. Katz, and A.R. Newton. Protection and versioning for oct. In *Proc. 26th ACM/IEEE Design Automation Conference*, pages 264–269, Las Vegas, June 1989.

[Shi81]   D.W. Shipman. The functional data model and the data language daplex. *ACM Trans. on Database Systems*, 6(1):140–173, March 1981.

[Sid80]   T.W. Sidle. Weaknesses of commercial data base management systems in engineering applications. In *Proc. 17th ACM/IEEE Design Automation Conference*, pages 57–61, June 1980.

[Sim93]   M.N. Sim. *Tool and Database Interfacing based on Virtual Objects*. PhD thesis, Delft University of Technology, Delft, Jan 1993.

[SS77]   J.M. Smith and D.C.P. Smith. Database abstractions: Aggregation and generalization. *ACM Trans. on Database Systems*, 2(2):105–133, June 1977.

[SW92]   G. Scholz and W. Wilkes. Information modelling of folded and unfolded design. In *Proc. EURO-DAC 92*, pages 459–464, Hamburg, Germany, Sept 1992.

[SZ89]   E. Siepmann and G. Zimmermann. An object-oriented datamodel for the vlsi design system playout. In *Proc. 26th ACM/IEEE Design Automation Conference*, pages 814–817, Las Vegas, June 1989.

[tB88]   J.H. ter Bekke. *Database Design*. Stenfert Kroese, Leiden, 1988. in Dutch.

[tB91]   J.H. ter Bekke. *Semantic Data Modeling in Relational Environments*. PhD thesis, Delft University of Technology, 1991. ISBN 90-9004132-X.

[tB92]   J.H. ter Bekke. *Semantic Data Modeling*. Prentice Hall, Englewood Cliffs, N.J., 1992. ISBN 0-13-806050-9.

[tBBvdW91] K.O. ten Bosch, P. Bingley, and P. van der Wolf. Design flow management in the nelsis cad framework. In *Proc. 28th ACM/IEEE Design Automation Conference*, pages 711–716, San Francisco, June 1991.

[tBvdWB93] K.O. ten Bosch, P. van der Wolf, and P. Bingley. A flow-based user interface for efficient execution of the design cycle. In *Proc. IEEE/ACM International Conference on CAD - 93*, pages 356–363, Santa Clara, Nov 1993.

[TL82]   D.C. Tsichritzis and F.H. Lochovsky. *Data Models*. Prentice-Hall, Englewood Cliffs, NJ, 1982.

[Tur85]   R. Turner. Cae data base plays key role in integration. *Computer Design*, pages 131–137, June 1985.

[Val90]   Validframe: High performance framework for electronic design, white paper. Valid Logic Systems, 1990. Valid Marketing Materials.

[vdH91]   P. van den Hamer. Focusing the big picture, a proposal for cfi's architecture-related activities. Technical report, Philips, Engineering Data Management Department, Jan 1991.

[vdH92]   A. van der Hoeven. *Concepts and Implementation of a Design System for Digital Signal Processor Arrays*. PhD thesis, Delft University of Technology, Delft, Oct 1992. ISBN 90-6275-816-9.

[vdHT90]   P. van den Hamer and M.A. Treffers. A data flow based architecture for cad frameworks. In *Proc. ICCAD - 90*, pages 482–485, 1990.

[vdHtBB+91]   P. van den Hamer, K.O. ten Bosch, P. Bingley, M.A. Treffers, and P. van der Wolf. A comparison of two approaches to design flow management by data schema analysis. Technical report, Philips Research Laboratories & Delft University of Technology, June 1991. JCF Project Deliverable, SP1.

[vdMvLvdW+87] N. van der Meijs, T.G.R. van Leuken, P. van der Wolf, I. Widya, and P. Dewilde. A data management interface to facilitate cad/ic software exchanges. In *Proc. IEEE ICCD '87*, pages 403–406, 1987.

[vdW86]   P. van der Wolf. Conceptual design of a design data management system for vlsi design. Master's thesis, Delft University of Technology, Delft, June 1986.

[vdWBD90]   P. van der Wolf, P. Bingley, and P. Dewilde. On the architecture of a cad framework: The nelsis approach. In *Proc. European Design Automation Conference*, pages 29–33, Glasgow, Scotland, March 1990.

[vdWSBD90]   P. van der Wolf, G.W. Sloof, P. Bingley, and P. Dewilde. Meta data management in the nelsis cad framework. In *Proc. 27th ACM/IEEE Design Automation Conference*, pages 142–145, Orlando, June 1990.

[vdWvL88]   P. van der Wolf and T.G.R. van Leuken. Object type oriented data modeling for vlsi data management. In *Proc. 25th ACM/IEEE Design Automation Conference*, pages 351–356, Anaheim, June 1988.

[vLvdW85]   T.G.R. van Leuken and P. van der Wolf. The icd design management system. In *Proc. IEEE ICCAD - 85*, pages 18–20, 1985.

[VMT92]   V. Vasudevan, Y. Mathys, and J. Tolar. Damocles: An observer-based approach to design tracking. In *Proc. IEEE/ACM International Conference on CAD - 92*, pages 546–551, Santa Clara, Nov 1992.

[Wat87]   M.L. Watkins. Software architecture and the unix operat-
               ing system: An introduction to interprocess communica-
               tion. *Hewlett-Packard Journal*, pages 26–36, June 1987.

[WBS82]   G. Wiederhold, A.F. Beetem, and G.E. Short. A database
               approach to communication in vlsi design. *IEEE Trans. on
               CAD of Integrated Circuits and Systems*, CAD-1(2):57–
               63, April 1982.

[WvLvdW88]   I. Widya, T.G.R. van Leuken, and P. van der Wolf. Con-
               currency control in a vlsi design database. In *Proc. 25th
               ACM/IEEE Design Automation Conference*, pages 357–
               362, Anaheim, June 1988.

[Zin81]   G. Zintl. A codasyl cad data base system. In *Proc. 18th
               ACM/IEEE Design Automation Conference*, pages 589–
               594, 1981.

INDEX

DATE DUE

HIGHSMITH #45230

Printed
in USA